Samuel William Johnson

How Crops Feed

A Treatise on the Atmosphere and the Soil as Related to the Nutrition...

Samuel William Johnson

How Crops Feed

A Treatise on the Atmosphere and the Soil as Related to the Nutrition...

ISBN/EAN: 9783744756471

Printed in Europe, USA, Canada, Australia, Japan

Cover: Foto ©berggeist007 / pixelio.de

More available books at **www.hansebooks.com**

HOW CROPS FEED.

A TREATISE ON THE

ATMOSPHERE AND THE SOIL

AS RELATED TO THE

Nutrition of Agricultural Plants.

WITH ILLUSTRATIONS.

BY

SAMUEL W. JOHNSON, M.A.,

PROFESSOR OF ANALYTICAL AND AGRICULTURAL CHEMISTRY IN THE SHEFFIELD
SCIENTIFIC SCHOOL OF YALE COLLEGE; CHEMIST TO THE CONNEC-
TICUT STATE AGRICULTURAL SOCIETY; MEMBER OF THE
NATIONAL ACADEMY OF SCIENCES.

NEW YORK:
ORANGE JUDD COMPANY,
1910

Entered according to Act of Congress, in the year 1870, by
ORANGE JUDD & CO.,
In the Clerk's Office of the District Court of the United States for the Southern District of New York

PRINTED IN U. S. A.

PREFACE.

The work entitled "How Crops Grow" has been received with favor beyond its merits, not only in America, but in Europe. It has been republished in England under the joint Editorship of Professors Church and Dyer, of the Royal Agricultural College, at Cirencester, and a translation into German is soon to appear, at the instigation of Professor von Liebig.

The Author, therefore, puts forth this volume—the companion and complement to the former—with the hope that it also will be welcomed by those who appreciate the scientific aspects of Agriculture, and are persuaded that a true Theory is the surest guide to a successful Practice.

The writer does not flatter himself that he has produced a popular book. He has not sought to excite the imagination with high-wrought pictures of overflowing fertility as the immediate result of scientific discussion or experiment, nor has he attempted to make a show of revolutionizing his subject by bold or striking speculations. His office has been to digest the cumbrous mass of evidence, in which the truths of Vegetable Nutrition lie buried out of the reach of the ordinary inquirer, and to set them forth in proper order and in plain dress for their legitimate and sober uses.

It has cost the Investigator severe study and labor to discover the laws and many of the facts which are laid down in the following pages. It has cost the Author no little work to collect and arrange the facts, and develop their mutual bearings, and the Reader must pay a similar price if he would apprehend them in their true significance.

In this, as in the preceding volume, the Author's method has been to bring forth all accessible facts, to present their evidence on the topics under discussion, and dispassionately to record their verdict. If this procedure be sometimes tedious, it is always safe, and there is no other mode of treating a subject which can satisfy the earnest inquirer.

It is, then, to the Students of Agriculture, whether on the Farm or in the School, that the Author commends his book, in confidence of receiving full sympathy for its spirit, whatever may be the defects in its execution.

CONTENTS.

Introduction. 17

DIVISION I.

THE ATMOSPHERE AS RELATED TO VEGETATION.

CHAPTER I.

ATMOSPHERIC AIR AS THE FOOD OF PLANTS.

§ 1. Chemical Composition of the Atmosphere..............................21
§ 2. Relation of Oxygen Gas to Vegetable Nutrition.........................22
§ 3.   "   " Nitrogen Gas to   "   " 26
§ 4.   "   " Atmospheric Water to Vegetable Nutrition..................34
§ 5.   "   " Carbonic Acid Gas "   " 38
§ 6.   '   " Atmospheric Ammonia to "   " 49
§ 7. Ozone..63
§ 8. Compounds of Nitrogen and Oxygen in the Atmosphere.................70
§ 9. Other Ingredients of the Atmosphere..................................91
§ 10. Recapitulation of the Atmospheric Supplies of Food to Crops...........94
§ 11. Assimilation of Atmospheric Food.....................................97
§ 12. Tabular View of the Relations of the Atmospheric Ingredients to the Life of Plants..98

CHAPTER II.

THE ATMOSPHERE AS PHYSICALLY RELATED TO VEGETATION.

§ 1. Manner of Absorption of Gaseous Food by Plants.........................99

DIVISION II.

THE SOIL AS RELATED TO VEGETABLE PRODUCTION.

CHAPTER I.

Introductory...104

CHAPTER II.

ORIGIN AND FORMATION OF SOILS.................................106

§ 1. Chemical Elements of Rocks........................107
§ 2. Mineralogical Elements of Rocks 108
§ 3. Rocks, their Kinds and Characters.................117
§ 4. Conversion of Rocks into Soil 122
§ 5. Incorporation of Organic Matter with the Soil, and its Effects..........135

CHAPTER III.

KINDS OF SOILS, THEIR DEFINITION AND CLASSIFICATION.

§ 1. Distinctions of Soils based upon the Mode of their Formation or Deposition..142
§ 2. Distinctions of Soils based upon Obvious or External Characters.......146

CHAPTER IV.

PHYSICAL CHARACTERS OF THE SOIL.........................157

§ 1. Weight of Soils....................................158
§ 2. State of Division.................................159
§ 3. Absorption of Vapor of Water.....................161
§ 4. Condensation of Gases............................165
§ 5. Power of Removing Solid Matters from Solution...171
§ 6. Permeability to Liquid Water. Imbibition. Capillary Power..........176
§ 7. Changes of Bulk by Drying and Frost..............183
§ 8. Adhesiveness.....................................184
§ 9. Relations to Heat................................186

CHAPTER V.

THE SOIL AS A SOURCE OF FOOD TO CROPS: INGREDIENTS WHOSE ELEMENTS ARE OF ATMOSPHERIC ORIGIN.

§ 1. The Free Water of the Soil in its Relations to Vegetable Nutrition......199
§ 2. The Air of the Soil..............................217
§ 3. Non-nitrogenous Organic Matters. Humus..........222
§ 4. The Ammonia of the Soil.........................238
§ 5. Nitric Acid (Nitrates) of the Soil..............251
§ 6. Nitrogenous Organic Matters of the Soil. Available Nitrogen..........274
§ 7. Decay of Organic Matters........................289
§ 8. Nitrogenous Principles of Urine.................293
§ 9. Comparative Nutritive Value of Ammonia-Salts and Nitrates..........300

CHAPTER VI.

THE SOIL AS A SOURCE OF FOOD TO CROPS: INGREDIENTS WHOSE ELEMENTS ARE DERIVED FROM ROCKS.

§ 1. General View of the Constitution of the Soil as Related to Vegetable Nutrition 305
§ 2. Aqueous Solution of the Soil 309
§ 3. Solution of the Soil in Strong Acids 329
§ 4. Portion of Soil Insoluble in Acids 330
§ 5. Reactions by which the Solubility of the Elements of the Soil is altered. Solvent Effects of Various Substances. Absorptive and Fixing Power of Soils 331
§ 6. Review and Conclusion 361

INDEX.

Absorption and displacement, law of.................336
Absorptive power of soils.........333
"   "   " cause of.343, 351
"   "   " significance of.........374
Acids in soil...................223
" absorbed by soils............355
Adhesion........................165
Adhesiveness of soils..............184
Air, atmospheric, composition of...21
" within the plant, composition of......................45
Alkali-salts, solvent effect of......130
Allotropism......................66
Alluvium........................145
Aluminum, alumina................107
Amides..........................276
Amide-like bodies...........277, 300
Ammonia.....................49, 54
" absorbed by clay....243, 267
"   "   " peat.....360
"   "   " plants....56, 98
" condensed in soils 240
" conversion of into nitric acid..................85
" evolved from flesh decaying under charcoal.....169
" fixed by gypsum........244
" in atmosphere...........54
" "   " how formed.77, 85
" of rain, etc...............60
" of the soil, formation of.239
" "   " chemically combined.243
" "   " physically condensed.240
" "   " quantity of..248
" "   " solubility of.246
" "   " volatility of..244

Ammonia-salts and nitrates, nutritive value of.................300
Amphibole.......................112
Analysis of soils, chemical indications of...............368
" mechanical...............147
Apatite.........................116
Apocrenates....................231
Apocrenic acid............227, 229
Argillite........................119
Ash-ingredients, quantity needful for crops..................363
Atmosphere, chemical composition................21, 22
" physical constitution..99
Atmospheric food, absorbed.......99
"   "   " assimilated....97
Barley crop, ash ingredients of.....364
Basalt.........................120
Bases, absorbed by soils......335, 359
Bisulphide of iron...............115
Burning of clay.................185
Calcite.........................115
Capillarity...........175, 199, 201
Carbohydrates in soil.............222
Carbon, fixed by plants.......43, 48
" in decay..................291
" supply of................95
Carbonate of lime.........115, 162
"   " magnesia........115, 162
Carbonic acid....................38
"   " absorbed by soil......221
"   "   " plants.41, 45, 98
"   " exhaled by plants..43, 99
"   " in the soil...........218
"   "   " water of soil..220
"   " quantity in the air.40, 47, 94
"   " solubility in water.40, 130

11

Carbonic acid, solvent action of....128
Carbonic oxide............... .. 92
Chabasite..........................115
" action on saline compounds.................345
" formed in Roman masonry.....................351
Chalk soils.........................192
Charcoal, absorbs gases.............165
" defecating action of......174
Chili saltpeter.....................253
Chlorite...........................113
Chrysolite.........................114
Clay.....................132, 134, 154
" absorptive power...........174
" effect of, on urine............293
Clay-slate.........................119
Coffee, condenses gases............168
Color of soil......................190
Conglomerates.....................121
Crenates..........................231
Crenic acid...................227, 229
Decay.............................289
Deliquescence.....................163
Deserts...........................197
Dew...........................189, 195
Diffusion of gases.................100
Diorite............................120
Dolerite...........................120
Dolomite......................115, 121
Draining..........................185
Drain water, composition of.......312
Drift..............................144
Dye stuffs, fixing of................174
Earth-closet......................171
Eremacausis......................289
Evaporation, produces cold........188
" amount of, from soil..197
Exhalation....................202, 206
Exposure of soil...................195
Feldspar..........................108
" growth of barley in........160
Fermentation......................290
Fixation of bases in the soil......339
Frost, effects of, on rocks.........124
" " " on soils......184, 185
Fumic acid.........................258
Gases, absorbed by the plant......103
" " " porous bodies..167
" " " soils......165, 166
" diffusion of...............100
" osmose of.................102
Glaciers...........................124

Glycine...........................296
Glycocoll.........................296
Gneiss............................119
Granite......................118, 120
Gravel.............................152
" warmth of..................195
Guanin............................296
Gypsum...........................115
" does not directly absorb water..................163
" fixes ammonia..............244
Hardpan...........................156
Heat, absorption and radiation of..
188, 193
" developed in flowering......24
" of soil......................187
Hippuric acid..................295, 277
Hornblende.......................112
Hydration of minerals.............127
Hydraulic cement.................122
Hydrochloric acid gas............. 93
Hydrogen, supply of, to plants..... 95
" in decay...............291
Hydrous silicates, formation of....352
Hygroscopic quality...............164
Humates..........................230
Humic acid...................226, 229
Humin........................236, 229
Humus..................136, 224, 276
" absorbent power for water..162
" absorbs salts from solutions.172
" action on minerals..........138
" chemical nature of..........138
" does it feed the plant?......232
" not essential to crops........238
" value of....................182
Iodine in sea-water................322
Isomorphism......................111
Kreatin...........................196
Kaolinite....................113, 132
Latent heat.......................188
Lawes' and Gilbert's wheat experiments........................372
Leucite...........................113
Lime, effects of..............184, 185
Limestone....................121, 122
Loam.............................154
Lysimeter........................314
Magnesite.........................115
Marble............................121
Marl..............................155
Marsh gas.....................91, 99
Mica..............................109

Mica slate........................119
Minerals..................106, 108
" hydration of............127
" solution of..............127
" variable composition of....110
Moisture, effect of, on temperature
  of soil......................195
Mold............................156
Moor-bed pan....................157
Muck............................155
Nitrate of ammonia......... 71, 73
" " " in atmosphere. 89
Nitrates........................252
" as food for plants..........271
" formed in soil........171, 179
" in water....................270
" loss of.....................270
" reduction of......... 73, 82, 85
" " " in soil........268
" tests for................... 75
Nitric acid...................... 70
" " as plant-food........ 90, 98
" " deportment towards the
    soil 357
" " in atmosphere........... 86
" " " rain-water............ 86
" " " soil............251, 254
" " " " sources of........256
Nitric oxide.................... 72
Nitric peroxide................. 72
Nitrification..............252, 286
" conditions of......265, 292
Nitrogen, atmospheric supply to
    plants................. 95
" combined, in decay.291, 292
" " of the soil...275
" combined, of the soil,
    available..............283
" combined, of the soil,
    inert..................278
" combined, of the soil,
    quantity needed for crops.288
" free, absorbed by soil...167
" " assimilated by the
    soil 259
" " in soil..............218
" " not absorbed by
    vegetation....26, 99
" " not emitted by liv-
    ing plants........ 23
Nitrogen-compounds, formation of,
    in atmosphere.75, 77, 83

Nitrogenous fertilizers, effect on
    cereals... 83
Nitrogenous organic matters of soil.274
Nitrous acid 72
Nitrous oxide...............71, 93
Ocher...........................156
Oxidation, aided by porous bodies.
    169, 170
Oxide of iron, a carrier of oxygen..257
" " hydrated, in the soil.350
Oxygen, absorbed by plants........ 98
" essential to growth...... 23
" exhaled by foliage......25, 99
" function of, in growth..... 24
" in soil....................218
" supply..................... 91
" weathering action of..... 131
Ozone............................ 63
" concerned in oxidation of ni-
    trogen.................... 82
" formed by chemical action.66, 67
" produced by vegetation.67, 84, 99
" relations of, to vegetable nu-
    trition..................... 70
Pan, composition of.............352
Parasitic plants, nourishment of...235
Peat......................155, 224
" nitrogen of...................274
Phosphate of lime...............116
Phosphoric acid fixed by the soil..357
" " presence in soil
    water.......... 315
Phosphorite.....................116
Plant-food, concentration of........320
" maintenance of supply.371
Platinized charcoal..............170
Platinum sponge, condenses oxygen170
Porphyry........................120
Potash, quantity in barley crop.....362
Provence, drouths of............198
Pyrites..........................115
Pyroxene........................112
Pumice..........................120
Putrefaction....................290
Quartz................... 108, 122
Rain-water, ammonia in........60, 88
" nitric acid in.......... 86
" phosphoric acid in.... 94
Ree Ree bottom, soil of.........160
Respiration of the plant............ 43
Rocks.......................106, 117
" attacked by plants............140

Rocks, conversion into soils........122
Roots, direct action on soil........326
Saline incrustations................179
Saltpeter..........................252
Salts decomposed or absorbed by the soil........................336
Sand...........................153, 162
Sand filter........................172
Sandstone..........................121
Serpentine.....................114, 121
Schist, micaceous..................119
" talcose......................121
" chlorite.....................121
Shales.............................122
Sherry wine region.................192
Shrinking of soils.................183
Silica.............................108
" function in the soil.........353
" of soil, liberated by strong acids........................330
Silicates..........................109
" zeolitic, presence in soils..349
Silicic acid, fixed in the soil....358
Silk, hygroscopic..................165
Soapstone..........................121
Sod, temperature of................199
Soil...............................101
" absorptive power of..........333
" aid to oxidation.............170
" aqueous solution of....309, 323, 328
" condenses gases..........165, 166
" capacity for heat............194
" chemical action in...........331
" composition of..........362, 369
" exhaustion of................373
" inert basis..................305
" natural strength of..........372
" origin and formation..106, 122, 135
" physical characters..........157
" porosity of..................176
" portion insoluble in acids...330
" relative value of ingredients....367
" reversion to rock............332
" solubility in acids..........329
" " water..................309
" source of food to crops......305
" state of division............159
Soils, sedentary...................143
" transported..................143
" weight of....................158
Solubility, standards of...........308

Solution of soil in acids..........370
" " " water............310
Steatite...........................121
Swamp muck.........................155
Sulphates, agents of oxidation.....258
Sulphate of lime...................115
Sulphur, in decay..................293
Sulphurous acid....................94
Sulphydric acid....................94
Syenite............................120
Talc...............................113
Temperature of soil.......186, 187, 194
Transpiration..................202, 208
Trap rock..........................120
Ulmates............................230
Ulmic acid.................224, 226, 229
Ulmin......................224, 226, 229
Urea..........................291, 277
Uric acid.....................295, 277
Urine..............................293
" preserved fresh by clay.......293
" its nitrogenous principles assimilated by plants..........296
Vegetation, antiquity of...........138
" decay of................137
" action on soil..........140
Volcanic rocks, conversion to soil..135
Wall fruits........................199
Water absorbed by roots.......202, 210
" functions of, in nutrition of plant....................216
" imbibed by soil..............180
" movements in soil............177
" proportion of in plant, influenced by soil..............213
" of soil..................315, 317
" " bottom water..........200
" capillary....................200
" hydrostatic..................199
" hygroscopic..................201
" quantity favorable to crops..214
Water-currents.....................124
Water-vapor, absorbed by soil.161, 164
" exhaled by plants......99
" not absorbed by plants..............35, 99
" of the atmosphere.....31
Weathering....................131-134
Wilting............................203
Wool, hygroscopic..................164
Zeolites......................114, 349

HOW CROPS FEED.

INTRODUCTION.

In his treatise entitled "How Crops Grow," the author has described in detail the Chemical Composition of Agricultural Plants, and has stated what substances are indispensable to their growth. In the same book is given an account of the apparatus and processes by which the plant takes up its food. The sources of the food of crops are, however, noticed there in but the briefest manner. The present work is exclusively occupied with the important and extended subject of Vegetable Nutrition, and is thus the complement of the first-mentioned treatise. Whatever information may be needed as preliminary, to an understanding of this book, the reader may find in "How Crops Grow." *

That crops grow by gathering and assimilating food is a conception with which all are familiar, but it is only by following the subject into its details that we can gain hints that shall apply usefully in Agricultural Practice.

---

* It has been at least the author's aim to make the first of this series of books prepare the way for the second, as both the first and the second are written to make possible an intelligible account of the mode of action of Tillage and of Fertilizers, which will be the subject of a third work.

When a seed germinates in a medium that is totally destitute of one or all the essential elements of the plant, the embryo attains a certain development from the materials of the seed itself (cotyledons or endosperm,) but shortly after these are consumed, the plantlet ceases to increase in dry weight,* and dies, or only grows at its own expense.

A similar seed deposited in ordinary soil, watered with rain or spring water and freely exposed to the atmosphere, evolves a seedling which survives the exhaustion of the cotyledons, and continues without cessation to grow, forming cellulose, oil, starch, and albumin, increases many times—a hundred or two hundred fold—in weight, runs normally through all the stages of vegetation, blossoms, and yields a dozen or a hundred new seeds, each as perfect as the original.

It is thus obvious that *Air*, *Water*, and *Soil*, are capable of feeding plants, and, under purely natural conditions, do exclusively nourish all vegetation.

In the soil, atmosphere, and water, can be found no trace of the peculiar organic principles of plants. We look there in vain for cellulose, starch, dextrin, oil, or albumin. The natural sources of the food of crops consist of various salts and gases which contain the ultimate elements of vegetation, but which require to be collected and worked over by the plant.

The embryo of the germinating seed, like the bud of a tree when aroused by the spring warmth from a dormant state, or like the sprout of a potato tuber, enlarges at the expense of previously organized matters, supplied to it by the contiguous parts.

As soon as the plantlet is weaned from the stores of the

---

* Since vegetable matter may contain a variable amount of water, either that which belongs to the sap of the fresh plant, or that which is hygroscopically retained in the pores, all comparisons must be made on the *dry*, i. e., *water-free* substance. See "How Crops Grow," pp. 53-5.

mother seed, the materials, as well as the mode of its nutrition, are for the most part completely changed. Henceforth the tissues of the plant and the cell-contents must be principally, and may be entirely, built up from purely inorganic or mineral matters.

In studying the nutrition of the plant in those stages of its growth that are subsequent to the exhaustion of the cotyledons, it is needful to investigate separately the nutritive functions of the Atmosphere and of the Soil, for the important reason that the atmosphere is nearly constant in its composition, and is beyond the reach of human influence, while the soil is infinitely variable and may be exhausted to the verge of unproductiveness or raised to the extreme of fertility by the arts of the cultivator.

In regard to the Atmosphere, we have to notice minutely the influence of each of its ingredients, including Water in the gaseous form, upon vegetable production.

The evidence has been given in "How Crops Grow," which establishes what fixed earthy and saline matters are essential ingredients of plants. The Soil is plainly the exclusive source of all those elements of vegetation which cannot assume the gaseous condition, and which therefore cannot exist in the atmosphere. The study of the soil involves a consideration of its origin and of its manner of formation. The productive soil commonly contains atmospheric elements, which are important to its fertility; the mode and extent of their incorporation with it are topics of extreme practical importance. We have then to examine the significance of its water, of its ammonia, and especially of its nitrates. These subjects have been recently submitted to extended investigations, and our treatise contains a large amount of information pertaining to them, which has never before appeared in any publication in the English tongue.

Those characters of the soil that indirectly affect the growth of plants are of the utmost moment to the farmer. It is through the soil that a supply of solar heat, without which no life is possible, is largely influenced. Water, whose excess or deficiency is as pernicious as its proper quantity is beneficial to crops, enters the plant almost exclusively through its roots, and hence those qualities of the soil which are most favorable to a due supply of this liquid demand careful attention. The absorbent power of soils for the elements of fertilizers is a subject which is treated of with considerable fullness, as it deserves.

Our book naturally falls into two divisions, the first of which is devoted to a discussion of the Relations of the Atmosphere to Vegetation, the second being a treatise on the Soil.

DIVISION I.

THE ATMOSPHERE AS RELATED TO VEGETATION.

CHAPTER I.

ATMOSPHERIC AIR AS THE FOOD OF PLANTS.

§ 1.

CHEMICAL COMPOSITION OF THE ATMOSPHERE.

A multitude of observations has demonstrated that from ninety-five to ninety-nine per cent of the entire mass (weight) of agricultural plants is derived directly or indirectly from the atmosphere.

The general composition of the Atmosphere is familiar to all. It is chiefly made up of the two elementary gases, Oxygen and Nitrogen, which have been described in "How Crops Grow," pp. 33–39.* These two bodies are present in the atmosphere in very nearly, though not altogether, invariable proportions. Disregarding its other ingredients, the atmosphere contains in 100 parts

	By weight.	By volume.
Oxygen	23.17	20.95
Nitrogen	76.83	79.05
	100.00	100.00

Besides the above elements, several other substances oc-

---

* In our frequent references to this book we shall employ the abbreviation H. C. G.

cur or may occur in the air in minute and variable quantities, viz.:

Water, as vapor	...average proportion by weight,				$1/100$	
Carbonic acid gas	"	"	"	"	$6/10{\cdot}000$	
Ammonia	"	"	"	"	$1/60{\cdot}000{\cdot}000$?	
Ozone	"	"	"	"	minute traces.	
Nitric acid	"	"	"	"	"	"
Nitrous acid	"	"	"	"	"	"
Marsh gas	"	"	"	"	"	"
In air of towns. { Carbonic oxide,	"	"	"	"	"	"
Sulphurous acid,	"	"	"	"	"	"
Sulphydric acid	"	"	"	"	"	"

Miller gives for the air of England the following average proportions by volume of the four most abundant ingredients.—(*Elements of Chemistry*, part II., p. 30, 3d Ed.)

```
Oxygen .................................20.61
Nitrogen ...............................77.95
Carbonic acid............................ .04
Water-vapor............................. 1.40
                                       ──────
                                       100.00
```

We may now appropriately proceed to notice in order each of the ingredients of the atmosphere in reference to the question of vegetable nutrition. This is a subject regarding which unaided observation can teach us little or nothing. The atmosphere is so intangible to the senses that, without some finer instruments of investigation, we should forever be in ignorance, even of the separate existence of its two principal elements. Chemistry has, however, set forth in a clear light many remarkable relations of the Atmosphere to the Plant, whose study forms one of the most instructive chapters of science.

§ 2.

RELATIONS OF OXYGEN GAS TO VEGETABLE NUTRITION.

Absorption of Oxygen Essential to Growth.—The element Oxygen is endowed with great chemical activity. This activity we find exhibited in the first act of vegeta tion, viz.: in germination. We know that the presence of oxygen is an indispensable requisite to the sprouting seed, and is possibly the means of provoking to action the dormant life of the germ. The ingenious experiments of Traube (H. C. G., p. 326.) demonstrate conclusively that free oxygen is an essential condition of the growth of the seedling plant, and must have access to the plumule, and especially to the parts that are in the act of elongation.

De Saussure long ago showed that oxygen is needful to the development of the buds of maturer plants. He experimented in the following manner: Several woody twigs (of willow, oak, apple, etc.) cut in spring-time just before the buds should unfold were placed under a bell-glass containing common air, as in fig. 1. Their cut extremities stood in water held in a small vessel, while the air of the bell was separated from the external atmosphere by the mercury contained in the large basin. Thus situated, the buds opened as in the free air, and oxygen gas was found to be consumed in considerable quantity. When, however, the twigs were confined in an atmosphere of nitrogen or hydrogen, they decayed, without giving any signs of vegetation. (*Recherches sur la Vegetation*, p. 115.)

Fig. 1.

The same acute investigator found that oxygen is absorbed by the roots of plants. Fig. 2 shows the arrangement by which he examined the effect of different gases on these organs. A young horse-chestnut plant, carefully lifted from the soil so as not to injure its roots, had the latter passed through the neck of a bell-glass, and the stem was then cemented air-tight into the opening. The bell was placed in a basin of mercury, C, D, to shut off its contents from the external air. So much water was introduced as to reach the ends of the principal roots, and the space above was occupied by common or some other kind of air. In one experiment carbonic acid, in a second nitrogen, in a third hydrogen, and in three others common air, was employed. In the first the roots died in seven or eight days, in the second and third they perished in thirteen or fourteen days, while in the three others they remained healthy to the end of three weeks, when the experiments were concluded. (*Recherches*, p. 104.)

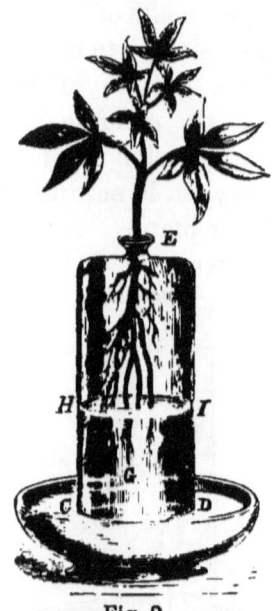

Fig. 2.

Flowers require oxygen for their development. Aquatic plants send their flower-buds above the water to blossom. De Saussure found that flowers consume, in 24 hours, several or many times their bulk of oxygen gas. This absorption proceeds most energetically in the pistils and stamens. Flowers of very rapid growth experience in this process, a considerable rise of temperature. Garreau, observing the spadix of *Arum italicum*, which absorbed $28\frac{1}{2}$ times its bulk of oxygen in one hour, found it 15° F. warmer than the surrounding air. In the ripening of fruits, oxygen is also absorbed in small quantity.

The Function of Free Oxygen.—All those processes of growth to which free oxygen gas is a requisite appear to depend upon the transfer to the growing organ of matters previously organized in some other part of the plant, and probably are not cases in which external inorganic bodies are built up into ingredients of the vegetable structure. Young seedlings, buds, flowers, and ripening fruits, have no power to increase in mass at the expense of the atmosphere and soil; they have no provision for the absorption of the nutritive elements that surround them externally, but grow at the expense of other parts of the plant (or seed) to which they belong. The function of free gaseous oxygen in vegetable nutrition, so far as can be judged from our existing knowledge, consists in effecting or aiding to effect the conversion of the materials which the leaves organize or which the roots absorb, into the proper tissues of the growing parts. Free oxygen is thus probably an agent of assimilation. Certain it is that the free oxygen which is absorbed by the plant, or, at least, a corresponding quantity, is evolved again, either in the uncombined state or in union with carbon as carbonic acid.

Exhalation of Oxygen from Foliage.—The relation of the *leaves* and *green parts* of plants to oxygen gas has thus far been purposely left unnoticed. These organs likewise absorb oxygen, and require its presence in the atmosphere, or, if aquatic, in the water which surrounds them; but they also, *during their exposure to light, exhale oxygen.*

Fig. 3.

This interesting fact is illustrated by a simple experiment. Fill a glass funnel with any kind of fresh leaves, and place it, inverted, in a wide glass containing water, fig. 3, so that it shall be completely immersed, and displace all air from its interior by agitation. Close the neck of the funnel air-tight by a cork, and pour off a portion of the water from the outer vessel. Expose now the leaves to strong sunlight. Observe that very soon minute bubbles of air will gather on the leaves. These will gradually increase in size and detach themselves, and after an hour or two, enough gas will accumulate in the neck of the funnel to enable the experimenter to prove that it consists of oxygen. For this purpose bring the water outside the neck to a level with that inside; have ready a splinter of pine, the end of which is glowing hot, but not in flame, remove the cork, and insert the ignited stick into the gas. It will inflame and burn much more brightly than in the external air. (See H. C. G., p. 35, Exp. 5.) To this phenomenon, one of the most important connected with our subject, we shall recur under the head of carbonic acid, the compound which is the chief source of this exhaled oxygen.

§ 3.

RELATIONS OF NITROGEN GAS TO VEGETABLE NUTRITION.

Nitrogen Gas not a Food to the Plant.—Nitrogen in the free state appears to be indifferent to vegetation. Priestley, to whom we are much indebted for our knowledge of the atmosphere, was led to believe in 1779 that free nitrogen is absorbed by and feeds the plant. But this philosopher had no adequate means of investigating the subject. De Saussure, twenty years later, having command of better methods of analyzing gaseous mixtures, concluded from his experiments that free nitrogen does not at all participate in vegetable nutrition.

Boussingault's Experiments.—The question rested until 1837, when Boussingault made some trials, which, however, were not decisive. In 1851–1855 this ingenious chemist resumed the study of the subject and conducted a large number of experiments with the greatest care, all of which lead to the conclusion that no appreciable amount of free nitrogen is assimilated by plants.

His plan of experiment was simply to cause plants to grow in circumstances where, every other condition of development being supplied, the only source of nitrogen at their command, besides that contained in the seed itself, should be the free nitrogen of the atmosphere. For this purpose he prepared a soil consisting of pumice-stone and the ashes of stable-manure, which was perfectly freed from all *compounds of nitrogen* by treatment with acids and intense heat. In nine of his earlier experiments the soil thus prepared was placed at the bottom of a large glass globe, *B*, fig. 4, of 15 to 20 gallons' capacity. Seeds of cress, dwarf beans, or lupins, were deposited in the soil, and a proper supply of water, purified for the purpose, was added. After germination of the seeds, a glass globe, *D*, of about one-tenth the capacity of the larger vessel, was filled with carbonic acid (to supply carbon), and was secured airtight to the mouth of the latter, communication being had between them by the open neck at *C*. The apparatus was then disposed in a suitably lighted place in a garden, and left to itself for a period which varied in the different experiments from 1½ to 5 months. At the conclusion of the trial the plants were lifted out, and, together with the soil from which their roots could not be entirely separated, were subjected to chemical analysis, to determine the amount of nitrogen which they had assimilated during growth.

Fig. 4.

The details of these trials are contained in the subjoined

Table. The weights are expressed in the gram and its fractions.

Number.	Kind of Plant.	Duration of Experiment.	Number of Seeds.	Weight of Seeds.	Weight of Crop.	Nitrogen in Seeds.	Nitrogen in Crop and Soil.	Gain (+) or Loss (—) of Nitrogen.
1	Dwarf bean.	2 months	1	0.780	1.87	0.0349	0.0340	—0.0009
2	Oat	2 "	10	0.377	0.54	0.0078	0.0067	—0.0011
3	Bean	3 "	1	0.530	0.89	0.0210	0.0189	—0.0021
4	"	3 "	1	0.618	1.13	0.0245	0.0226	—0.0019
5	Oat	2½ "	4	0.139	0.44	0.0031	0.0030	—0.0001
6	Lupin	1½ "	2	0.825	1.82	0.0480	0.0483	+0.0003
7	"	2 "	6	2.202	6.73	0.1282	0.1246	—0.0036
8	"	7 weeks	2	0.600	1.95	0.0349	0.0339	—0.0010
9	"	6 "	1	0.343	1.05	0.0200	0.0204	+0.0004
10	"	6 "	2	0.686	1.53	0.0399	0.0397	—0.0002
11	Dwarf bean	2 months	1	0.792	2.35	0.0354	0.0360	+0.0006
12	" "	2½ "	1	0.665	2.80	0.0298	0.0277	—0.0021
13	{ Cress	3½ "	3	0.008	0.65	0.0013	0.0013	0.0000
	{ "	as manure	10	0.026				
14	{ Lupin	5 months	2	0.627	5.76	0.1827	0.1697	—0.0130
	{ "	as manure	8	2.512				
14	Sum			11.720	30.11	0.6135	0.5868	—0.0247

While it must be admitted that the unavoidable errors of experiment are relatively large in working with such small quantities of material as Boussingault here employed, we cannot deny that the aggregate result of these trials is decisive against the assimilation of free nitrogen, since there was a loss of nitrogen in the 14 experiments, amounting to 4 per cent of the total contained in the seeds; while a gain was indicated in but 3 trials, and was but 0.13 per cent of the nitrogen concerned in them.—(Boussingault's *Agronomie, Chimie Agricole, et Physiologie*, Tome I, pp. 1-64.)

The Opposite Conclusions of Ville.—In the years 1849, '50, '51, and '52, Georges Ville, at Grenelle, near Paris, experimented upon the question of the assimilability of free nitrogen. His method was similar to that first employed by Boussingault. The plants subjected to his trials were cress, lupins, colza, wheat, rye, maize, sun-flowers, and tobacco. They were situated in a large octagonal cage made of iron sashes, set with glass-plates. The air was constantly renewed, and carbonic acid was introduced in proper quantity. The experiments were conducted on a larger scale than those of Boussingault, and their result was uniformly the reverse. Ville indeed thought to have established that vegetation feeds on the free nitrogen of the air. To the conclusions to which Boussingault drew from the trials made in the manner already described, Ville objected that the limited amount of air contained in the glass globes was insufficient for the needs of vegetation; that plants, in fact, could not attain a normal development under the conditions of Boussingault's experiments.— (Ville, *Recherches sur la Vegetation*, pp. 29–58, and 53–98.)

Boussingault's Later Experiments.—The latter thereupon instituted a new series of trials in 1854, in which he proved that the plants he had previously experimented upon attain their full development in a confined atmosphere under the circumstances of his first experiments, provided they are supplied with some assimilable *compound of nitrogen*. He also conducted seven new experiments in an apparatus which allowed the air to be constantly renewed, and in every instance confirmed his former results.— (*Agronomie, Chimie Agricole et Physiologie*, Tome I, pp. 65–114.)

The details of these experiments are given in the following Table. The weights are expressed in grams.

Number.	Kind of Plant.	Duration of Experiment.	Number of Seeds.	Weight of Seeds.	Weight of Crop.	Nitrogen in Seeds.	Nitrogen in Crop.	Gain (+) or Loss (−) of Nitrogen.
1	Lupin............	10 weeks	1	0.337	2.140	0.0196	0.0187	−0.0009
2	Bean............	10 "	1	0.720	2.000	0.0322	0.0325	+0.0003
3	Bean............	12 "	1	0.748	2.847	0.0335	0.0341	+0.0006
4	Bean............	14 "	1	0.755	2.240	0.0339	0.0329	−0.0010
5	Bean............	13 "	2	1.510	5.150	0.0676	0.0666	−0.0010
6	Lupin............	9 as manure	1 1	0.310 0.300	1.730	0.0355	0.0324	−0.0021
7	Cress............	10 weeks as manure	30 12	0.100	0.533	0.0046	0.0052	+0.0006
			Sum	.4.780	16.64	0.2269	0.2240	−0.0035

Inaccuracy of Ville's Results.—In comparing the investigations of Boussingault and Ville as detailed in their own words, the critical reader cannot fail to be struck with the greater simplicity of the apparatus used by the former, and his more exhaustive study of the possible sources of error incidental to the investigation—facts which are greatly in favor of the conclusions of this skillful and experienced philosopher. Furthermore Cloëz, who was employed by a Commission of the French Academy to oversee the repetition of Ville's experiments, found that a considerable quantity of ammonia was either generated within or introduced into the apparatus of Ville during the period of the trials, which of course vitiated all his results.

Any further doubts with regard to this important subject have been effectually disposed of by another most elaborate investigation.

Research of Lawes, Gilbert, and Pugh.—In 1857 and '58, the late Dr. Pugh, afterward President of the Pennsylvania Agricultural College, associated himself with Messrs. Lawes and Gilbert, of Rothamstead, England, for the purpose of investigating all those points connected with the subject, which the spirited discussion of the researches of Boussingault and Ville had suggested as possibly accounting for the diversity of their results. Lawes, Gilbert, and Pugh, conducted 27 experiments on graminaceous and leguminous plants, and on buckwheat. The plants vegetated within large glass bells. They were cut off from the external air by the bells dipping into mercury. They were supplied with renewed portions of purified air mixed with carbonic acid, which, being forced into the bells instead of being drawn through them, effectually prevented any ordinary air from getting access to the plants.

To give an idea of the mode in which these delicate investigations are conducted, we give here a figure and concise description of the appara-

ATMOSPHERIC AIR AS THE FOOD OF PLANTS.

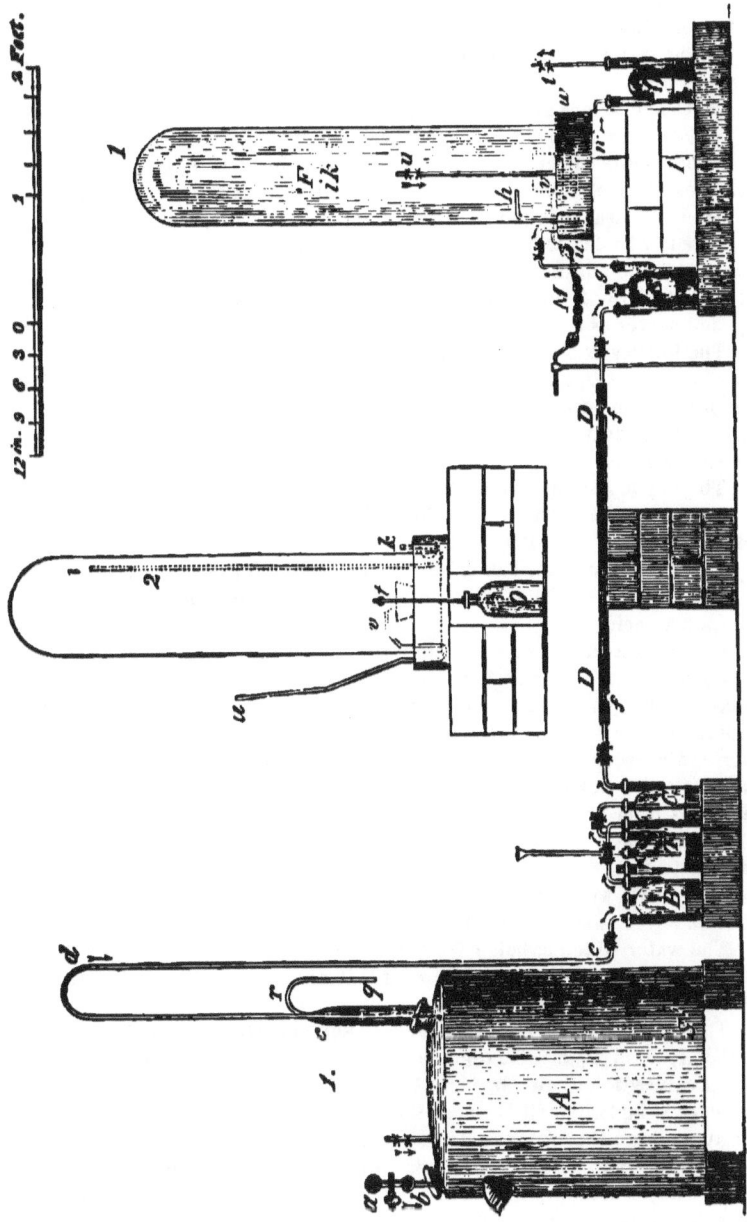

Fig. 5.

tus employed by Lawes, Gilbert, and Pugh, in their experiments made in the year 1858.

A, fig. 5, represents a stone-ware bottle 18 inches in diameter and 24 inches high.

B, *C*, and *E*, are glass 3-necked bottles of about 1 quart capacity.

F is a large glass shade 9 inches in diameter and 40 inches high.

a represents the cross-section of a leaden pipe, which, passing over all the vessels *A* of the series of 16, supplied them with water, from a reservoir not shown, through the tube with stop-cock *a b*.

c d e is a leaden exit-tube for air. At *c* it widens, until it enters the vessel *A*, and another bent tube, *q r s*, passes through it and reaches to the bottom of *A*, as indicated by the dotted lines. The latter opens at *q*, and serves as a safety tube to prevent water passing into *d e*.

The bottles *B C* are partly filled with strong sulphuric acid.

The tube *D D*, 1 inch wide and 3 feet long, is filled with fragments of pumice-stone saturated with sulphuric acid. At *f f* indentations are made to prevent the acid from draining against the corks with which the tube is stopped.

The bottle *E* contains a saturated solution of pure carbonate of soda.

g h is a bent and caoutchouc-jointed glass tube connecting the interior of the bottle *E* with that of the glass shade *F*.

i k, better indicated in 2, is the exit-tube for air, connecting the interior of the shade *F* with an eight-bulbed apparatus, *M*, containing sulphuric acid.

w w is a vessel of glazed stone-ware, containing mercury in a circular groove, into which the lower edge of the shade *F* is dipped. These glass tubes, *g h*, *u v*, and *i k*, 2, pass under the edge of the shade and communicate with its interior, the mercury cutting off all access of exterior air, except through the tubes. Another tube, *n o*, passes air-tight through the bottom of the stone-ware vessel, and thus communicates with its interior.

The tubes *u v* and *i k* are seen best in 2, which is taken at right angles to 1.

The plants were sprouted and grew in pots, *v*, within the shades. The tube *u v* was to supply them with water.

The water which exhaled from the foliage and gathered on the inside of the shade ran off through *n o* into the bottle *O*. This water was returned to the pots through *u v*.

The renewed supply of pure air was kept up through the bottles and tube *A*, *B*, *C*, *D*, *E*. On opening the cock *a b*, *A*, water enters *A*, and its pressure forces air through the bottles and tube into the shade *F*, whence it finds its exit through the tube *i k*, and the bulb-apparatus *M*. In its passage through the strong sulphuric acid of *B*, *C*, and *D*, the air is completely freed from ammonia, while the carbonate of soda of *E* removes any traces of nitric acid. The sulphuric acid of the bulb M purifies the small amount of air that might sometimes enter the shade through the tube *i k*, owing to cooling of the air in *F*, when the current was not passing. The outer ends of the tubes l and u were closed with caoutchouc tubes and glass plugs.

In these experiments it was considered advisable to furnish to the plants more carbonic acid than the air contains. This was accomplished by pouring hydrochloric acid from time to time into the bottle T, which contained fragments of marble. The carbonic acid gas thus liberated joined, and was swept on by the current of air in C. Experiments taught how much hydrochloric acid to add and how often. The proportion of this gas was kept within the limits which previous experimenters had found permissible, and was not allowed to exceed 4.0 per cent. nor to fall below 0.2 per cent.

In these experiments the seeds were deposited in a soil purified from nitrogen-compounds, by calcination in a current of air and subsequent washing with pure water. To this soil was added about 0.5 per cent of the ash of the plant which was to grow in it. The water used for watering the plants was specially purified from ammonia and nitric acid.

The experiments of Lawes, Gilbert, and Pugh, fully confirmed those of Boussingault. For the numerous details and the full discussion of collateral points bearing on the study of this question, we must refer to their elaborate memoir, "On the Sources of the Nitrogen of Vegetation."—(*Philosophical Transactions*, 1861, II, pp. 431-579.)

Nitrogen Gas is not Emitted by Living Plants.—It was long supposed by vegetable physiologists that when the foliage of plants is exposed to the sun, free nitrogen is evolved by them in small quantity. In fact, when plants are placed in the circumstances which admit of collecting the gases that exhale from them under the action of light, it is found that besides oxygen a quantity of gas appears, which, unless special precautions are observed, consists chiefly of nitrogen, which was a part of the air that fills the intercellular spaces of the plant, or was dissolved in the water, in which, for the purposes of experiment, the plant is immersed.

If, as Boussingault has recently (1863) done, this air be removed from the plant and water, or rather if its quantity be accurately determined and deducted from that obtained in the experiment, the result is that no nitrogen gas remains. A small quantity of gas besides oxygen was indeed usually evolved from the plant when submerged in water. The gas on examination proved to be marsh gas.

Cloëz was unable to find marsh gas in the air exhaled from either aquatic or land plants submerged in water, and in his most recent researches (1865) Boussingault found none in the gases given off from the foliage of a living tree examined without submergence.

The ancient conclusion of Saussure, Daubeny, Draper, and others, that nitrogen is emitted from the substance of the plant, is thus shown to have been based on an inaccurate method of investigation.

§ 4.

RELATIONS OF ATMOSPHERIC WATER TO VEGETABLE NUTRITION.

Occurrence of Water in the Atmosphere.—If water be exposed to the air in a shallow, open vessel for some time, it is seen to decrease in quantity, and finally disappears entirely; it evaporates, vaporizes, or volatilizes. It is converted into vapor. It assumes the form of air, and becomes a part of the atmosphere.

The rapidity of evaporation is greater the more elevated the temperature of the water, and the drier the atmosphere that is over it. Even snow and ice slowly suffer loss of weight in a dry day though it be frosty.

In this manner evaporation is almost constantly going on from the surface of the ocean and all other bodies of water, so that the air always carries a portion of aqueous vapor.

On the other hand, a body or mixture whose temperature is far lower than that of the atmosphere, condenses vapor from the air and makes it manifest in the form of water. Thus a glass of ice-water in a warm summer's day becomes externally bedewed with moisture. In a similar manner, dew deposits in clear and calm summer nights upon the surface of the ground, upon grass, and upon all exposed objects, whose temperature rapidly falls when they cease to be warmed by the sun. Again, when the invisible vapor which fills a hot tea-kettle or steam-boiler issues into cold air, a visible cloud is immediately formed, which consists of minute droplets of water. In like manner, fogs and the clouds of the sky are produced by the cooling of air charged with vapor. When the cooling is sufficiently great and sudden, the droplets acquire such size as to fall directly to the ground; the water assumes the form of rain.

Water then exists in the atmosphere during the periods of vegetable activity as gas or vapor,* and as liquid. In the former state it is almost perpetually rising into the air, while in the latter form it frequently falls again to the ground. It is thus in a continual transition, back and forth, from the earth to the sky, and from the sky to the earth.

We have given the average quantity of water-vapor in the air at one per cent; but the amount is very variable, and is almost constantly fluctuating. It may range from less than one-half to two and a half or three per cent, according to temperature and other circumstances.

When the air is damp, it is saturated with moisture, so that water is readily deposited upon cool objects. On the other hand, when dry, it is capable of taking up additional moisture, and thus facilitates evaporation.

Is Atmospheric Water Absorbed by Plants?—It has long been supposed that growing vegetation has the power to absorb vapor of water from the atmosphere by its foliage, as well as to imbibe the liquid water which in the form of rain and dew may come in contact with its leaves. Experiments which have been instituted for the purpose of ascertaining the exact state of this question have, however, demonstrated that agricultural plants gather little or no water from these sources.

The wilting of a plant results from the fact that the leaves suffer water to evaporate from them more rapidly than the roots can take it up. The speedy reviving of a wilted plant on the falling of a sudden rain or on the deposition of dew depends, not so much on the absorption by the foliage, of the water that gathers on it, as it does

---

* While there is properly no essential difference between a gas and a vapor, the former term is commonly applied more especially to aëriform bodies which are not readily brought to the liquid state, and the latter to those which are easily condensed to liquids or solids.

on the suppression of evaporation, which is a consequence of the saturation of the surrounding air with water.

Unger, and more recently Duchartre, have found, 1st, that plants lose weight (from loss of water) in air that is as nearly as possible saturated with vapor, when their roots are not in contact with soil or liquid water. Duchartre has shown, 2d, that plants do not gain, but sometimes lose weight when their foliage only is exposed to dew or even to rain continued through eighteen hours, although they increase in weight strikingly (from absorption of water through their roots,) when the rain is allowed to fall upon the soil in which they are planted.

Knop has shown, on the other hand, that leaves, either separate or attached to twigs, gain weight by continued *immersion* in water, and not only recover what they may have lost by exposure, but absorb more than they originally contained. (*Versuchs-Stationen*, VI, 252.)

The water of dews and rains, it must be remembered, however, does not often thoroughly wet the absorbent surface of the leaves of most plants; its contact being prevented, to a great degree, by the hairs or wax of the epidermis.

Finally, 3d, Sachs has found that even the roots of plants appear incapable of taking up watery vapor.

To convey an idea of the method employed in such investigations, we may quote Sachs' account of one of his experiments. (*V. St.*, II, 7.) A young camellia, having several fresh leaves, was taken from the loose soil of the pot in which it had been growing; from its long roots all particles of earth were carefully removed, and its weight was ascertained. The bottom of a glass cylinder was covered with water to a little depth, and the roots of the camellia were introduced, but not in contact with the water. The stem was supported at its lower part in a hole in a glass cover,* that was cemented air-tight upon the vessel. The stem itself was cemented by soft wax into the hole, so that the interior of the vessel was completely cut off from direct communication with the external atmosphere. The plant thus situated had its roots in an atmosphere as nearly as possible saturated with vapor of water, while its leaves were exposed to the external air. After four days had expired, the entire apparatus, plant included, had lost 1.823 grm. Thereupon the plant was removed from the vessel and weighed by itself; it had lost 2.188 grm. The loss of the entire apparatus was due to vapor of water, which had escaped through the leaves. The difference between this loss and the loss which the plant had experienced could be attributed only to an exhalation of water through the roots, and amounted to (2.188 — 1.823=) 0.365 grm.

This exhalation of water into the confined and moist atmosphere of the glass vessel is explained, according to Sachs, by the fact that the chemical changes proceeding within the plant elevate its temperature above that of the surrounding atmosphere.

Knop, in experiments on the transpiration of plants, (*V. St.*, VI, 255,) obtained similar results. He found, however, that a moist piece of paper or wood also lost weight when kept for some time in a confined space over water. He therefore concludes that it is nearly impossible in the conditions of such experiments to maintain the air saturated with vapor, and that the loss of weight by the roots is due, not to the heat arising from internal chemical changes, but to simple evaporation from their surface. In one instance he found that a portulacca standing over night in a bell-glass with moistened sides, did not lose, but gained weight, some dew having gathered on its foliage.

---

* The cover consisted of two semicircular pieces of ground glass, each of which had a small semicircular notch, so that the two could be brought together by their straight edges around the stem.

The result of these investigations is, that while, perhaps, wilted foliage in a heavy rain may take up a small quantity of water, and while foliage and roots may absorb some vapor, yet in general and for the most part the atmospheric water is not directly taken up to any great extent by plants, and does not therefore contribute immediately to their nourishment.

Atmospheric Water Enters Crops through the Soil.—It is only after the water of the atmosphere has become incorporated with the soil, that it enters freely into agricultural plants. The relations of this substance to proper vegetable nutrition may then be most appropriately discussed in detail when we come to consider the soil. (See p. 199.)

It is probable that certain air-plants (epiphytes) native to the tropics, which have no connection with the soil, and are not rooted in a medium capable of yielding water, condense vapor from the air in considerable quantity. So also it is proved that the mosses and lichens absorb water largely from moist air, and it is well known that they become dry and brittle in hot weather, recovering their freshness and flexibility when the air is damp.

§ 5.

RELATIONS OF CARBONIC ACID GAS TO VEGETABLE NUTRITION.

Composition and Properties of Carbonic Acid.—When 12 grains of pure carbon are heated to redness in 32 grains of pure oxygen gas, the two bodies unite together, themselves completely disappearing, and 44 grains of a gas are produced which has the same bulk as the oxygen had at the beginning of the experiment. The new gas is nearly one-half heavier than oxygen, and differs in most of its properties from both of its ingredients. It is carbonic acid. This substance is the product of the burning of charcoal in oxygen gas, (H. C. G., p. 35, Exp. 6.) It is, in fact, produced whenever any organic body is burned or decays in contact with the air. It is like oxygen, colorless, but it has a peculiar pungent odor and pleasant acid taste.

The composition of carbonic acid is evident from what has been said as to its production from carbon and oxygen. It consists of two atoms, or 32 parts by weight, of oxygen, united to one atom, or 12 parts, of carbon. Its symbol is CO_2. In the subjoined scheme are given its symbolic, atomic, and percentage composition.

	At. wt.	Per cent.
C =	12	27.27
O_2 =	32	72.73
CO_2 =	44	100.00

In a state of combination carbonic acid exists in nature in immense quantities. Limestone, marble, and chalk, contain, when pure, 44 per cent of this acid united to lime. These minerals are in chemical language carbonate of lime. Common salæratus is a carbonate of potash, and soda-salæratus is a carbonate of soda.

From either of these carbonates it is easy to separate this gas by the addition of another and stronger acid.

For this purpose we may employ the Rochelle or Seidlitz powders so commonly used in medicine. If we mingle together in the dry state the contents of a blue paper, which contains carbonate of soda, with those of a white paper, which consist of tartaric acid, nothing is observed. If, however, the mixture be placed at the bottom of a tall bottle, and a little water be poured upon it, at once a vigorous bubbling sets in, which is caused by the liberated carbonic acid.*

Some important properties of the gas thus set free may be readily made manifest by the following experiments.

a. If a burning taper or match be immersed in the gas, the flame is immediately extinguished. This happens because of the absence of free oxygen.

b. If the mouth of the bottle from which carbonic acid is escaping be held to that of another bottle, the gas can be poured into the second vessel, on account of its density being one-half greater than that of the air. Proof that the invisible gas has thus been transferred is had by placing

---

* Chalk, marble, or salæratus, and chlorhydric (muriatic) acid, or strong vinegar (acetic acid) can be equally well employed.

a burning taper in the second bottle, when, if the experiment was rightly conducted, the flame will be extinguished.

c. Into a bottle filled as in the last experiment with carbonic acid, some lime-water is poured and agitated. The previously clear lime-water immediately becomes turbid and milky from the formation of *carbonate of lime*, which is nearly insoluble in water.

Carbonic Acid in the Atmosphere.—To show the presence of carbonic acid in the atmosphere, it is only necessary to expose lime-water in an open vessel. But a little time elapses before the liquid is covered with a white film of carbonate. As already stated, the average proportion of carbonic acid in the atmosphere is 6–10000ths (1–1600th nearly) by weight, or 4–10000ths (1–2500th) by bulk. Its quantity varies somewhat, however. Among over 300 analyses made by De Saussure in Switzerland, Verver in Holland, Lewy in New Granada, and Gilm in Austria, the extreme range was from 47 to 86 parts by weight in 100,000.

Deportment of Carbonic Acid towards Water.—Water dissolves carbonic acid to a greater or less extent, according to the temperature and pressure. Under the best ordinary conditions it takes up about its own volume of the gas. At the freezing point it may absorb nearly twice as much. This gas is therefore usually found in spring, well, and river waters, as well as in dew and rain. The considerable amount held in solution in cold springs and wells is a principal reason of the refreshing quality of their water. Under pressure the proportion of carbonic acid absorbed by water is much larger, and when the pressure is removed, a portion of the gas escapes, resuming its gaseous form and causing effervescence. The liquid that flows from a soda-fountain is an aqueous solution of carbonic acid, made under pressure. Bottled cider, ale, champagne, and all effervescent beverages, owe their sparkle and much of their refreshing qualities to the carbonic acid they contain.

The Absorption of Carbonic Acid by Plants.—In 1771 Priestley, in England, found that the leaves of plants immersed in water, sometimes disengaged carbonic acid, sometimes oxygen, and sometimes no gas at all. A few years later Ingenhouss proved that the exhalation of carbonic acid takes place in the absence, and that of oxygen in the presence, of solar light. Several years more elapsed before Sennebier first demonstrated that the oxygen which is exhaled by foliage in the sunlight comes from the carbonic acid contained in the water in which the plants are immersed for the purpose of these experiments. It had been already noticed, by Ingenhouss, that in spring water plants evolve more oxygen than in river water. We now know that the former contains more carbonic acid than the latter. Where the water is by accident or purposely free from carbonic acid, no gas is evolved by foliage in the sunlight.

The attention of scientific men was greatly attracted by these interesting discoveries; and shortly Percival, in England, found that a plant of mint whose roots were stationed in water, flourished better when the air bathing its foliage was artificially enriched in carbonic acid than in the ordinary atmosphere.

In 1840 Boussingault furnished direct proof, of what indeed was hardly to be doubted, viz.: the absorption of the carbonic acid of the atmosphere by foliage.

Into one of the orifices in a three-necked glass globe he introduced and fixed air-tight the branch of a living vine bearing twenty leaves; with another opening he connected a tube through which a slow current of air, containing, in one experiment, four-10000ths of carbonic acid, could be passed into the globe. This air after streaming over the vine leaves, at the rate of about 15 gallons per hour, escaped by the third neck into an arrangement for collecting and weighing the carbonic acid that remained in it. The experiment being set in process in the sunlight, it was found that the enclosed foliage removed from the current of air *three-fourths* of the carbonic acid it at first contained.

Influence of the Relative Quantity of Carbonic Acid.—De Saussure investigated the influence of various proportions of carbonic acid mixed with atmospheric air on the development of vegetation. He found that young peas (4 inches high) when exposed to direct sunlight, endured for some days an atmosphere consisting to one-half of carbonic acid. When the proportion of this gas was increased to two-thirds or more, they speedily withered. In air containing one-twelfth of carbonic acid the peas flourished much better than in ordinary atmospheric air. The average increase of each of the plants exposed to the latter for five or six hours daily during ten days was eight grains; while in the former it amounted in the same time to eleven grains. In the shade, however, Saussure found that increase of the proportion of carbonic acid to one-twelfth was detrimental to the plants. Their growth under these circumstances was but three-fifths of that experienced by similar plants exposed to the same light for the same time, but in common air. He also proved that foliage cannot long exist in the total absence of carbonic acid, when exposed to *direct sunlight.* This result was obtained by enclosing young plants whose roots were immersed in water, or the branches of trees stationed in the soil, in a vessel which contained moistened quicklime. This substance rapidly absorbs and fixes carbonic acid, forming carbonate of lime. Thus situated, the leaves began in a few days to turn yellow, and in two to three weeks they dropped off.

In darkness the presence of lime not only did not destroy the plants, but they prospered the better for its presence, i. e., for the absence or constant removal of carbonic acid.

Boussingault has lately shown that *pure* carbonic acid is decomposed by leaves exposed to sunlight with extreme slowness, or not at all. It must be mixed with some other gas, and when diluted with either oxygen, nitrogen, or hydrogen, or even when rarefied by the air-pump to a certain extent, the absorption and decomposition proceed as usual.

Conclusion. — It thus is proved 1st, that vegetation

can flourish only when its foliage is bathed by an atmosphere which contains a certain small amount of carbonic acid; 2d, that this gas is absorbed by the leaves, and, under the influence of sunlight, is decomposed within the plant, its carbon being retained, and in an unknown manner becoming a part of the plant itself, while the oxygen is exhaled into the atmosphere in the free state.

Relative volumes of absorbed Carbonic Acid and exhaled Oxygen.—From the numerous experiments of De Saussure, and from similar ones made recently with greatly improved means of research by Unger and Knop, it is established that in sunlight the volume of oxygen exhaled is nearly equal to the volume of carbonic acid absorbed. Since free oxygen occupies the same bulk as the carbonic acid produced by uniting it with carbon, it is evident that carbon mainly and not oxygen to much extent, is retained by the plant from this source.

Respiration and Fixation of Carbon by Plants.—In 1851 Garreau, and in 1858 Corenwinder, reviewed experimentally the whole subject of the relations of plants to carbonic acid. Their researches fully confirm the conclusions derived from older investigations, and furnish some additional facts.

We have already seen (p. 22) that the plant requires free oxygen, and that this gas is absorbed by those parts of vegetation which are in the act of growth. As a consequence of this entrance of oxygen into the plant, a corresponding amount of carbonic acid is produced within and exhales from it. There go on accordingly, in the expanding plant, two opposite processes, viz., the absorption of oxygen and exhalation of carbonic acid, and the absorption of carbonic acid and evolution of oxygen. The first process is chemically analogous with the breathing of animals, and may hence be designated as *respiration*. We may speak of the other process as the *fixation of carbon*.

These opposite changes obviously cannot take place at the same points, but must proceed in different organs or cells, or in different parts of the same cells. They furthermore tend to counterbalance each other in their effects on the atmosphere surrounding the plant. The processes to which the absorption of oxygen and evolution of carbonic acid are necessary, appear to go on at all hours of the day and night, and to be independent of the solar light. The production of carbonic acid is then continually occurring; but, under the influence of the sun's direct rays, the opposite absorption of carbonic acid and evolution of oxygen proceed so much more rapidly, that when we experiment with the entire plant the first result is completely masked. In our experiments we can, in fact, only measure the preponderance of the latter process over the former. In sunlight it may easily happen that the carbonic acid which exhales from one cell is instantly absorbed by another, and likewise the oxygen, which escapes from the latter, may be in part imbibed by the former.

In total darkness it is believed that carbonic acid is not absorbed and decomposed by the plant, but only produced in, and exhaled from it. In no case has any evolution of oxygen been observed in the absence of light.

When, instead of being exposed to the direct rays of the sun, only the diffused light of cloudy days or the softened light of a dense forest acts upon them, plants may, according to circumstances, exhale either oxygen or carbonic acid in preponderating quantity. In his earlier investigations, Corenwinder observed an exhalation of carbonic acid in diffused light in the cases of tobacco, sunflower, lupine, cabbage, and nettle. On the contrary, he found that lettuce, the pea, violet, fuchsia, periwinkle, and others, evolved oxygen under similar conditions. In one instance a bean exhaled neither gas. These differences are not peculiar to the plants just specified, but depend upon the intensity of the light and the stage of development in which the plant exists. Corenwinder noticed that the evolution of carbonic acid in diffused light was best exhibited by very young plants, and mostly ceased as they grew older.

Corenwinder has confirmed and extended these observations in more recent investigations. (*Ann. d. Sci. Nat.*, 1864, I, 297.)

He finds that buds and young leaves exhale carbonic acid (and absorb oxygen) by day, even in bright sunshine. He also finds that all leaves exhale carbonic acid not alone at night, but likewise by day, when placed in the diffused light of a room, illuminated from only one side. A plant, which in full light yields no carbonic acid to a slow stream of air passing its foliage, immediately gives off the gas when carried into such an apartment, and *vice versa*.

Amount of Carbonic Acid absorbed.—The quantity of carbonic acid absorbed by day in direct light is vastly greater than that exhaled during the night. According to Corenwinder's experiments, 15 to 20 minutes of direct sunlight enable colza, the pea, the raspberry, the bean, and sunflower, to absorb as much carbonic acid as they exhale during a whole night.

As to the amount of carbonic acid whose carbon is retained, Corenwinder found that a single colza plant took up in one day of strong sunshine more than two quarts of the gas.

Boussingault (*Comptes Rend.*, Oct. 23d, 1865) found as the average of a number of experiments, that a square meter of oleander leaves decomposed in sunlight 1.108 liters of carbonic acid per hour. In the dark, the same surface of leaf exhaled but 0.07 liter of this gas.

Composition of the Air within the Plant.—Full confirmation of the statements above made is furnished by tracing the changes which take place within the vegetable tissues. Lawes, Gilbert, and Pugh, (*Phil. Trans.*, 1861, II, p. 486,) have examined the composition of the air contained in plants, as well when the latter are removed from, as when they are subjected to, the action of light. To collect the gas from the plants, the latter were placed in a glass vessel filled with water, from which all air had been expelled by long boiling and subsequent cooling in full and tightly closed bottles. The vessel was then connected with a simple apparatus in which a vacuum was produced by the fall of mercury, down a tube of 30 inches height. The air contained within the cells of the plant was thus drawn over into the vacuum and collected for examination. We give some of the results of the 6th series of their examinations. "The Table shows the Amount and Composition of the Gas evolved into a Torricellian vacuum by duplicate portions of oat-plant, both kept in the dark for some time, and then one exposed to sunlight for about twenty minutes, when both were submitted to exhaustion."

			Per cent.		
Date, 1858.	Conditions during Exhaustion.	Cubic centimeters of Gas collected.	Nitrogen.	Oxygen.	Carbonic Acid.
July 31.	In dark.	24.0	77.08	3.75	19.17
	In sunlight.	34.5	68.69	24.93	6.38
Aug. 2.	In dark.	10.6	68.28	10.21	21.51
	In sunlight.	39.2	67.86	25.95	6.89
Aug. 2.	In dark.	30.7	76.87	8.14	14.99
	In sunlight.	26.5	69.43	27.17	3.40

These analyses show plainly what it is that happens in the cells of the plant. The atmospheric air freely penetrates the vegetable tissues, (II. C. G., p. 288.) In darkness, the oxygen that is thus contained within the plant takes carbon from the vegetable matter and forms with it carbonic acid. This process goes on with comparative rapidity, and the proportion of oxygen may be diminished from 21, the normal percentage, to 4, or even, as in some other experiments, to less than 1 per cent of the volume of the air. Upon bringing the vegetable tissue into sunlight, the carbonic acid previously formed within the cells undergoes decomposition. with separation of its oxygen in the free gaseous condition, while its carbon remains in the solid state as a constituent of the plant. Referring to the table above, we see that twenty minutes' exposure to the solar rays was sufficient in the second experiment (where the proportion of nitrogen remained nearly unaltered) to decompose 14 per cent of carbonic acid and liberate its oxygen. The total volume of air collected was 2.4 cubic inches, and the volume of decomposed carbonic acid was $\frac{1}{8}$ of a cubic inch, that of the liberated oxygen being the same.

Supply of Carbonic Acid in the Atmosphere.—Although this body forms but $\frac{1}{10000}$ of the weight of the atmosphere, yet such is the immense volume of the latter that it is calculated to contain, when taken to its entire height, no less than 3,400,000,000,000 tons of carbonic acid. This amounts to about 28 tons over every acre of the earth's surface.

According to Chevandier, an acre of beech-forest annually assimilates about one ton (1950 lbs.) of carbon, an amount equivalent to $3\frac{1}{2}$ tons of this gas. Were the whole earth covered with this kind of forest, and did it depend solely upon the atmosphere for carbon, eight years must elapse before the existing supply would be exhausted, in case no means had been provided for restoring to the air what vegetation constantly removes.

When we consider that but one-fourth of the earth's surface is land, and that on this the annual vegetable production is very far below (not one-third) the amount stated above for thrifty forest, we are warranted in assuming the atmospheric content of carbonic acid sufficient, without renewal, for a hundred years of growth. This ingredient of the atmosphere is maintained in undiminished quantity by the oxidation of carbon in the slow decay of organic matters, in the combustion of fuel, and in animal respiration.

That the carbonic acid of the atmosphere may fully suffice to provide a rapidly growing vegetation with carbon is demonstrated by numerous facts. Here we need only mention that in a soil totally destitute of all carbon, besides that contained in the seeds sown in it, Boussingault brought sunflowers to a normal development. The writer has done the same with buckwheat; and Sachs, Knop, Stohmann, Nobbe and Siegert, and others, have produced perfect plants of maize, oats, etc., whose roots, throughout the whole period of growth, were immersed in a weak, saline solution, destitute of carbon. (See II. C. G., *Water Culture*, p. 167.)

Hellriegel's recent experiments give the result that the atmospheric supply of carbonic acid is probably sufficient for the production of a maximum crop under all circumstances; at least artificial supply, whether of the gas, of its aqueous solution, or of a carbonate, to the soil, had no effect to increase the crop. (*Chem. Ackersmann*, 1868, p. 18.)

Liebig considers carbonic acid to be, under all circumstances, the exclusive source of the carbon of agricultural vegetation. To this point we shall recur in our study of the soil.

Carbon fixed by Chlorophyll.—The fixation of carbon from the carbonic acid of the air is accomplished in, or has an intimate relation with, the chlorophyll grains of the leaf or green stem. This is not only evident from the microscopic study of the development of the carbohydrates, especially starch, whose organization proceeds from the chlorophyll, but is an inference from the experiments of Gris on the effects of withholding iron from plants. In absence of iron, the leaf may unfold and attain a certain development; but chlorophyll is not formed, and the plant soon dies, without any real growth by assimilation of food from without. (II. C. G., p. 200.) Finally, experiment shows that oxygen is given off (and carbonic acid decomposed with fixation of carbon) only from those parts of plants in which the microscope reveals chlorophyll, although the prevailing color may be other than green.

Influence of Light on Fixation of Carbon.—As mentioned, Ingenhouss (in 1779) discovered that oxygen gas is given off from foliage, and carbon fixed in the plant only under the influence of light. Experiments show that when a seed germinates in exclusion of light it not only does not gain, but steadily loses weight from the consumption of carbon (and hydrogen) in slow oxidation (respiration).

Thus Boussingault (*Comptes Rendus*, 1864, p. 883) caused two beans to germinate and vegetate, one in the ordinary light and one in darkness, during 26 days. The gain in light and loss in darkness in entire (dry) weight, and of carbon, etc., are seen from the statement below.

	In Light.	*In Darkness.*
Weight of seed	0.922 gram	0.926 gram.
Weight of plant	1.293 "	0.566 "
Gain =	0.371 gram.	Loss....0.360 gram.
Carbon, Gain =	0.1926 "	Loss...0.1598 "
Hydrogen, " =	0.0200 "	" ...0.0232 "
Oxygen, " =	0.1591 "	" ...0.1766 "

§ 6.

THE AMMONIA OF THE ATMOSPHERE AND ITS RELATIONS TO VEGETABLE NUTRITION.

Ammonia is a gas, colorless and invisible, but having a peculiar pungency of odor and an acrid taste.

Preparation.—It may be obtained in a state of purity by heating a mixture of chloride of ammonium (sal ammoniac) and quicklime. Equal quantities of the two substances just named (50 grams of each) are separately pulverized, introduced into a flask, and well mixed by shaking. A straight tube 8 inches long is now secured in the neck of the flask by means of a perforated cork, and heat applied. The ammonia gas which soon escapes in abundance is collected in dry bottles, which are inverted over the tube. The gas, rapidly entering the bottle, in a few moments displaces the twice heavier atmospheric air. As soon as a feather wet with vinegar or dilute chlorhydric acid becomes surrounded with a dense smoke when approached to the mouth of the bottle, the latter may be removed, corked, and another put in its place. Three or four pint bottles of gas thus collected will serve to illustrate its properties, as shortly to be noticed.

Solubility in Water.—This character of ammonia is exhibited by removing, under cold water, the stopper of a bottle filled with the gas. The water rushes with great violence into the bottle as into a vacuum, and entirely fills it, provided all atmospheric air had been displaced.

The *aqua ammonia*, or *spirits of hartshorn* of the druggist, is a strong solution of ammonia, prepared by passing a stream of ammonia gas into cold water. At the freezing point, water absorbs 1,150 times its bulk of ammonia. When such a solution is warmed, the gas escapes abundantly, so that, at ordinary summer temperatures, only one-half the ammonia is retained. If the solution be heated to boiling, all the ammonia is expelled before the water has nearly boiled away. The gas escapes even from very dilute solutions when they are exposed to the air, as is at once recognized by the sense of smell.

Composition.—When ammonia gas is heated to redness by being made to pass through an ignited tube, it is decomposed, loses its characteristic odor and other properties, and is resolved into a mixture of nitrogen and hydrogen gases. These elements exist in ammonia in the proportion of one part by bulk of nitrogen, to three parts of hydrogen, or by weight fourteen parts or one atom of nitrogen and three parts, or three atoms of hydrogen. The subjoined scheme exhibits the composition of ammonia, as expressed in symbols, atoms, and percentages.

Symbol.		At. w't.		Per cent.
N	=	14	..	82.39
H_3	=	3	..	17.61
NH_3	=	17	..	100.00

Formation of Ammonia.—1. When hydrogen and nitrogen gases are mingled together in the proportions to form ammonia, they do not combine either spontaneously or by aid of any means yet devised, but remain for an indefinite period as a mere mixture. The oft repeated assertion that *nascent* hydrogen, i. e., hydrogen at the moment of liberation from some combination, may unite with free nitrogen to form ammonia, has been completely refuted by the experiments of Will, (*Ann. Ch. u. Ph.*, XLV, 110.) The ammonia observed by older experimenters existed, ready formed, in the materials they operated with.

2. It appears from recent researches (of Boettger, Schönbein, and Zabelin) that ammonia is formed in minute quantity from atmospheric nitrogen in many cases of combustion, and is also generated when vapor of water and the air act upon each other in contact with certain organic matters, at a temperature of 120° to 160° F. To this subject we shall again recur. p. 77.

3. Ammonia may result from the reduction of nitrous and nitric acids, and from the action of alkalies and lime upon the albuminoids, gelatine, and other similar organic matters. To these modes of its formation we shall recur on subsequent pages.

4. Ammonia is most readily and abundantly formed from organic nitrogenous bodies; e. g., the albuminoids and similar substances, by decay or by dry distillation. It is supposed to have been called ammonia because one of its most common compounds (sal ammoniac) was first prepared by burning camels' dung near the temple of Jupiter Ammon in Libya, Asia Minor. The name hartshorn, or spirits of hartshorn, by which it is more commonly known, was adopted from the circumstance of its preparation by distilling the horns of the stag or hart.

The ammonia and ammoniacal salts of commerce (carbonate of ammonia, sal ammoniac, and sulphate of ammonia) are exclusively obtained from these sources.

When urine is allowed to become stale, it shortly smells

of ammonia, which copiously escapes in the form of carbonate, and may be separated by distillation.

When bones are heated in close vessels, as in the manufacture of bone-black or bone-char for sugar refining, the liquid product of the distillation is strongly charged with carbonate of ammonia.

Commercial ammonia is mostly derived, at present, from the distillation of bituminous coal, and is a bye-product of the manufacture of illuminating gas. The gases and vapors that issue from the gas-retort in which the coal is heated to redness, are washed by passing through water. This wash water is always found to contain a small quantity of ammonia, which may be cheaply utilized

The exhalations of volcanoes and fumeroles likewise contain ammonia, which is probably formed in a similar manner.

In the processes of combustion and decay the elements of the organic matters are thrown into new groupings, which are mostly simpler in composition than the original substances. A portion of nitrogen and a corresponding portion of hydrogen then associate themselves to form ammonia.

Ammonia is a Strong Alkaline Base.—Those bases which have in general the strongest affinity for acids, are potash, soda, and ammonia. These bodies are very similar in many of their most obvious characters, and are collectively denominated the alkalies. They are alike freely soluble in water, have a bitter, burning taste, alike corrode the skin and blister the tongue; and, united with acids, form the most permanent saline compounds, or salts.

Carbonate of Ammonia.—If a bottle be filled with carbonic acid, (by holding it inverted over a candle until the latter becomes extinguished when passed a little way into the bottle,) and its mouth be applied to that of a vessel containing ammonia gas, the two invisible airs at once combine to a solid salt, the carbonate of ammonia, which appears as a white cloud where its ingredients come in contact.

Carbonate of ammonia occurs in commerce under the name "salts of hartshorn," and with the addition of some perfume forms the contents of the so-called smelling-bottles. It rapidly vaporizes, exhaling the odor of ammonia very strongly, and is hence sometimes termed *sal volatile*. Like camphor, this salt passes from the solid state into that of invisible vapor, at ordinary temperatures, without assuming intermediately the liquid form.

In the atmosphere the quantity of carbonic acid greatly preponderates over that of the ammonia; hence it is impossible that the latter should exist in the free state, and we must assume that it occurs there chiefly in combination with carbonic acid. The carbonate of ammonia, whether solid or gaseous, is readily soluble in water, and like free ammonia it evaporates from its solution with the first portions of aqueous vapor, leaving the residual water relatively free from it.

In the guano-beds of Peru and Bolivia, carbonate of ammonia is sometimes found in the form of large transparent crystals, which, like the artificially-prepared salt, rapidly exhale away in vapor, if exposed to the air.

This salt, commonly called bicarbonate of ammonia, contains in addition to carbonic acid and ammonia, a portion of water, which is indispensable to its existence. Its composition is as follows:

Symbol.	At. w't.	Per cent.
NH_3	17	21.5
H_2O	18	22.8
CO_2	44	55.7
$NH_3 . H_2O . CO_2$	79	100.0

Tests for Ammonia.—*a*. If salts of ammonia are rubbed together with *slaked lime*, best with the addition of a few drops of water, the ammonia is liberated in the gaseous state, and betrays itself (1) by its characteristic *odor*; (2) by its *reaction* on moistened test-papers; and

(3) by giving rise to the formation of *white fumes*, when any object (*e. g.*, a glass rod) moistened with hydrochloric acid, is brought in contact with it. These fumes arise from the formation of solid ammoniacal salts produced by the contact of the gases.

b. Nessler's Test.—For the detection of exceedingly minute traces of ammonia, a reaction first pointed out by Nessler may be employed. Digest at a gentle heat 2 grammes of iodide of potassium, and 3 grammes of iodide of mercury, in 5 cub. cent. of water; add 20 cub. cent. of water, let the mixture stand for some time, then filter; add to the filtrate 30 cub. cent. of pure concentrated solution of potassa (1 : 4); and, should a precipitate form, filter again. If to this solution is added, in small quantity, a liquid containing ammonia or an ammonia-salt, a *reddish brown precipitate*, or with exceedingly small quantities of ammonia, a *yellow coloration* is produced from the formation of dimercurammonic iodide, $NHg_2 I.OH_2$.

c. Bohlig's Test.—According to Bohlig, chloride of mercury (corrosive sublimate) is the most sensitive reagent for ammonia, when in the free state or as carbonate. It gives a *white precipitate*, or in very dilute solutions (even when containing but $\frac{1}{200,000}$ of ammonia) a *white turbidity*, due to the separation of mercurammonic chloride, $NH_2 Hg.Cl$. In solutions of the salts of ammonia with other acids than carbonic, a clear solution of mixed carbonate of potassa and chloride of mercury must be employed, which is prepared by adding 10 drops of a solution of the purest carbonate of potassa, (1 of salt to 50 of water,) and 5 drops of a solution of chloride of mercury to 80 c. c. of water exempt from ammonia (such is the water of many springs, but ordinary distilled water rarely). This reagent may be kept in closed vessels for a time without change. If much more concentrated, oxide of mercury separates from it. By its use the ammonia salt is first converted into carbonate by double decomposition with the carbonate of potassa, and the further reaction proceeds as before mentioned.

Occurrence of Ammonia in the Atmosphere.—The existence of ammonia in the atmosphere was first noticed by De Saussure, and has been proved repeatedly by direct experiment. That the quantity is exceedingly minute has been equally well established.

Owing partly to the variable extent to which ammonia occurs in the atmosphere, but chiefly to the difficulty of collecting and estimating such small amounts, the statements of those who have experimented upon this subject are devoid of agreement.

We present here a tabulated view of the most trustworthy results hitherto published:

1,000,000,000 parts of atmospheric air contain of ammonia, according to
Graeger, at Mühlhausen, Germany, average, 333 parts.
Fresenius, " Wiesbaden,   "   "   133 "
Pierre,   " Caen, France, 1851-52, "   3500 "
   "   "   "   "   1852-53, "   500 "
Bineau,   " Lyons,   "   1852-53, "   250 "
   "   " Caluire,   "   " winter,   40 "
   "   "   "   "   " summer,   80 "
Ville,   " Paris,   "   1849-50, average, 24 "
   "   " Grenelle,   "   1851,   "   21 "

Graham has shown by experiment (Ville, *Recherches sur la Vegetation*, Paris, 1853, p. 5,) that a quantity of ammonia like that found by Fresenius is sufficient to be readily detected by its effect on a red litmus paper, which is not altered in the air. This demonstrates that the atmosphere where Graham experimented (London) contained less than $^{133}/_{10,000,000}$ths of ammonia in the state of bicarbonate. The experiments of Fresenius and of Gräger were made with comparatively small volumes of air, and those of the latter, as well as those of Pierre, and some of Bineau's, were made in the vicinity of dwellings, or even in cities, where the results might easily be influenced by local emanations. Bineau's results were obtained by a method scarcely admitting of much accuracy.

The investigations of Ville (*Recherches*, Paris, 1853,) are, perhaps, the most trustworthy, having been made on a large scale, and apparently with every precaution. We may accordingly assume that the average quantity of ammonia in the air is one part in fifty millions, although the amount is subject to considerable fluctuation.

From the circumstance that ammonia and its carbonate are so readily soluble in water, we should expect that in rainy weather the atmosphere would be washed of its ammonia; while after prolonged dry weather it would contain more than usual, since ammonia escapes from its solutions with the first portions of aqueous vapor.

The Absorption of Ammonia by Vegetation.—The general fact that ammonia in its compounds is appropriated by plants as food is most abundantly established. The salts of ammonia applied as manures in actual farm practice have produced the most striking effects in thousands of instances.

By watering potted plants with very dilute solutions of ammonia, their luxuriance is made to surpass by far that of similar plants, which grow in precisely the same conditions, save that they are supplied with simple water.

Ville has stated, 1851-2, that vegetation in conservatories may be remarkably promoted by impregnating the air with gaseous carbonate of ammonia. For this purpose lumps of the solid salt are so disposed on the heating apparatus of the green-house as to gradually vaporize, or vessels containing a mixture of quicklime and sal ammoniac may be employed. Care must be taken that the air does not contain at any time more than four ten-thousandths of its weight of the salt; otherwise the foliage of tender plants is injured. Like results were obtained by Petzholdt and Chlebodarow in 1852-3.

Absorption of Ammonia by Foliage.—Although such facts indicate that ammonia is directly absorbed by foliage, they fail to prove that the soil is not the medium through which the absorption really takes place. We remember that according to Unger and Duchartre water enters the higher plants almost exclusively by the roots, after it has been absorbed by the soil. To Peters and Sachs (*Chem. Ackersmann*, 6, 158) we owe an experiment which appears to demonstrate that ammonia, like carbonic acid, is imbibed by the leaves of plants. The figure represents the apparatus employed. It consisted of a glass bell, resting below,

Fig. 6.

air-tight, upon a glass plate, and having two glass tubes cemented into its neck above, as in fig. 6. Through an aperture in the centre of the glass plate the stem of the plant experimented on was introduced, so that its foliage should occupy the bell, while the roots were situated in a pot of earth beneath. Two young bean-plants, growing in river sand, were arranged, each in a separate apparatus, as in the figure, on June 19th, 1859, their stems being cemented tightly into the opening below, and through the tubes the foliage of each plant received daily the same quantities of moist atmospheric air mixed with 4-5 per cent of carbonic acid. One plant was supplied in addition with a quantity of carbonate of ammonia, which was introduced by causing the air that was forced into the bell to stream through a dilute solution of this salt. Both plants grew well, until the experiment was terminated, on the 11th of August, when it was found that the plant whose foliage was not supplied with carbonate of ammonia weighed, dry, 4.14 gm., while the other, which was supplied with the vapor of this salt, weighed, dry, 6.74 gms. The first plant had 20 full-sized leaves and 2 side shoots; the second had 40 leaves and 7 shoots, besides a much larger mass of roots. The first contained 0.106 gm. of nitrogen; the second, double that amount, 0.208 gm. Other trials on various plants failed from the difficulty of making them grow in the needful circumstances.

The absorption of ammonia by foliage does not appear, like that of carbonic acid, to depend upon the action of sunlight; but, as Mulder has remarked,* may go on at all times, especially since the juices of plants are very frequently more or less charged with acids which **directly unite chemically with ammonia.**

When absorbed, ammonia is chiefly applied by agricul-

---

* Chemie der Ackerkrume, Vol. 2, p. 211.

tural plants to the production of the albuminoids.* We measure the nutritive effect of ammonia salts applied as fertilizers by the amount of nitrogen which vegetation assimilates from them.

Effects of Ammonia on Vegetation. — The remarkable effect of carbonate of ammonia upon vegetation is well described by Ville. We know that most plants at a certain period of growth under ordinary circumstances cease to produce new branches and foliage, or to expand those already formed, and begin a new phase of development in providing for the perpetuation of the species by producing flowers and fruit. If, however, such plants are exposed to as much carbonate of ammonia gas as they are capable of enduring, at the time when flowers are beginning to form, these are often totally checked, and the activity of growth is transferred to stems and leaves, which assume a new vigor and multiply with extraordinary luxuriance. If flowers are formed, they are sterile, and yield no seed.

Another noticeable effect of ammonia—one, however, which it shares with other substances—is its power of deepening the color of the foliage of plants. This is an indication of increased vegetative activity and health, as a pale or yellow tint belongs to a sickly or ill-fed growth.

A third result is that not only the mass of vegetation is increased, but the relative proportion of nitrogen in it is heightened. This result was obtained in the experiment of Peters and Sachs just described. To adduce a single other instance, Ville found that grains of wheat, grown in pure air, contained 2.09 per cent of nitrogen, while those which were produced under the influence of ammonia contained 3.40 per cent.

---

* In tobacco, to the production of nicotine; in coffee, of caffeine; and in many other plants to analogous substances. Plants appear oftentimes to contain *small quantities* of ammonia salts and nitrates, as well as of asparagin, ($C_4 H_8 N_2 O_3$,) a substance first found in asparagus, and which is formed in many plants when they vegetate in exclusion of light.

Do Healthy Plants Exhale Ammonia?—The idea having been advanced that in the act of vegetation a loss of nitrogen may occur, possibly in the form of ammonia, Knop made an experiment with a water-plant, the *Typha latifolia*, a species of Cat-tail, to determine this point. The plant, growing undisturbed in a pond, was enclosed in a glass tube one and a half inches in diameter, and six feet long. The tube was tied to a stake driven for the purpose; its lower end reached a short distance below the surface of the water, while the upper end was covered air-tight with a cap of India rubber. This cap was penetrated by a narrow glass tube, which communicated with a vessel filled with splinters of glass, moistened with pure hydrochloric acid. As the large tube was placed over the plant, a narrow U-shaped tube was immersed in the water to half its length, so that one of its arms came within, and the other without, the former. To the outer extremity of the U-tube was attached an apparatus, for the perfect absorption of ammonia. By aspirating at the upper end of the long tube, a current of ammonia-free air was thus made to enter the bottom of the apparatus, stream upward along the plant, and pass through the tube of glass-splinters wet with hydrochloric acid. Were any ammonia evolved within the long tube, it would be collected by the acid last named. To guard against any ammonia that possibly might arise from decaying matters in the water, a thin stratum of oil was made to float on the water within the tube. Through this arrangement a slow stream of air was passed for fifty hours. At the expiration of that time the hydrochloric acid was examined for ammonia; but none was discovered. Our tests for ammonia are so delicate, that we may well assume that this gas is not exhaled by the *Typha latifolia*.

The statements to be found in early authors (Sprengel, Schübler, Johnston), to the effect that ammonia is exhaled by some plants, deserve further examination.

The *Chenopodium vulvaria* exhales from its foliage a body chemically related to ammonia, and that has been mistaken for it. This substance, known to the chemist as trimethylamine, is also contained in the flowers of *Cratægus oxycantha*, and is the cause of the detestable odor of these plants, which is that of putrid salt fish.* (Wicke, *Liebig's Ann.*, 124, p. 338.)

Certain fungi (toad-stools) emit trimethylamine, or some analogous compound. (Lehmann, *Sachs' Experimental Physiologie der Pflanzen*, p. 273, note.)

It is not impossible that ammonia, also, may be exhaled from these plants, but we have as yet no proof that such is the case.

Ammonia of the Atmospheric Waters.—The ammonia proper to the atmosphere has little effect upon plants through their foliage when they are sheltered from dew and rain. Such, at least, is the result of certain experiments.

Boussingault (*Agronomie, Chimie Agricole, et Physiologie*, T. I, p. 141) made ten distinct trials on lupins, beans, oats, wheat, and cress. The seeds were sown in a soil, and the plants were watered with water both exempt from nitrogen. The plants were shielded by glazed cases from rain and dew, but had full access of air. The result of the ten experiments taken together was as follows:

```
Weight of seeds............. 4.965 grm's.
   "    "  dry harvest........18.730   "
Nitrogen in harvest and soil.. .2499   "
   "       "  seeds ........... .2307   "
           Gain of nitrogen..... .0192 grm's = 7.6 per cent of the
                                               total quantity.
```

When rains fall, or dews deposit upon the surface of the

---

* Trimethylamine $C_3H_9N = N(CH_3)_3$ may be viewed as ammonia NH_3, in which the three atoms of hydrogen are replaced by three atoms of methyl CH_3. It is a gas like ammonia, and has its pungency, but accompanied with the odor of stale fish. It is prepared from herring pickle, and used in medicine under the name propylamine.

soil, or upon the foliage of a cultivated field, they bring down to the reach of vegetation in a given time a quantity of ammonia, far greater than what is diffused throughout the limited volume of air which contributes to the nourishment of plants. The solubility of carbonate of ammonia in water has already been mentioned. In a rain-fall we have the atmosphere actually washed to a great degree of its ammonia, so that nearly the entire quantity of this substance which exists between the clouds and the earth, or in that mass of atmosphere through which the rain passes, is gathered by the latter and accumulated within a small space.

Proportion of Ammonia in Rain-water, etc.—The proportion of ammonia* which the atmospheric waters thus collect and bring down upon the surface of the soil, or upon the foliage of plants, has been the subject of investigations by Boussingault, Bineau, Way, Knop, Bobiere, and Bretschneider. The general result of their accordant investigations is as follows: In rain-water the quantity of ammonia in the entire fall is very variable, ranging in the country from 1 to 33 parts in 10 million. In cities the amount is larger, tenfold the above quantities having been observed.

The first portions of rain that fall usually contain much more ammonia than the latter portions, for the reason that a certain amount of water suffices to wash the air, and what rain subsequently falls only dilutes the solution at first formed. In a long-continued rain, the water that finally falls is almost devoid of ammonia. In rains of short duration, as well as in dews and fogs, which occasionally are so heavy as to admit of collecting to a sufficient extent for analysis, the proportion of ammonia is greatest, and is the greater the longer the time that has elapsed since a previous precipitation of water.

---

* In all quantitative statements regarding ammonia, NH_3 is to be understood, and not NH_4O.

Boussingault found in the first tenth of a slow-falling rain (24th Sept., 1853) 66 parts of ammonia, in the last three-tenths but 13 parts, to 10 million of water. In dew he found 40 to 62; in fog, 25 to 72; and in one extraordinary instance 497 parts in ten million.

Boussingault found that the average proportion of ammonia in the atmospheric waters (dew and fogs included) which he was able to collect at Liebfrauenberg (near Strasburg, France) from the 26th of May to the 8th of Nov. 1853, was 6 parts in 10 million (*Agronomie*, etc., T. II, 238). Knop found in the rains, snow, and hail, that fell at Moeckern, near Leipzig, from April 18th to Jan. 15th, 1860, an average of 14 parts in 10 million. (*Versuchs-Stationen*, Vol. 3, p. 120.)

Pincus and Röllig obtained from the atmospheric waters collected at Insterburg, North Prussia, during the 12 months ending with March, 1865, in 26 analyses, an average of 7 parts of ammonia in 10 million of water. The average for the next following 12 months was 9 parts in 10 million.

Bretschneider found in the atmospheric waters collected by him at Ida-Marienhütte, in Silesia, from April, 1865, to April, 1866, as the average of 9 estimations, 30 parts of ammonia in 10 million of water. In the next year the quantity was 23 parts in 10 million.

In 10 million parts of rain-water, etc., collected at the following places in Prussia, were contained of ammonia—at Regenwalde, in 1865, 24; in 1867, 28; at Dahme, in 1865, 17; at Kuschen, in 1865, $5\frac{1}{2}$; and in 1866, $7\frac{1}{2}$ parts. (*Preus. Ann. d. Landwirthschaft*, 1867.) The monthly averages fluctuated without regularity, but mostly within narrow limits. Occasionally they fell to 2 or 3 parts, once to nothing, and rose to 35 or 40, and once to 144 parts in 10 million.

Quantity of Ammonia per Acre Brought Down by Rain, etc.—In 1855 and '56, Messrs. Lawes & Gilbert, at Rothamstead, England, collected on a large rain-gauge having

ATMOSPHERIC AIR AS THE FOOD OF PLANTS. 63

a surface of $\frac{1}{1000}$ of an acre, the entire rain-fall (dews, etc., included) for those years. Prof. Way, at that time chemist to the Royal Ag. Soc. of England, analyzed the waters, and found that the total amount of ammonia contained in them was equal to 7 lbs. in 1855, and 9½ lbs. in 1856, for an acre of surface. These amounts were yielded by 663,000 and 616,000 gallons of rain-water respectively.

In the waters gathered at Insterburg during the twelve-month ending March, 1865, Pincus and Röllig obtained 6.38 lbs. of ammonia per acre.

Bretschneider found in the waters collected at Ida-Marienhütte from April, 1865, to April, 1866, 12 lbs. of ammonia per acre of surface.

The significance of these quantities may be most appropriately discussed after we have noticed the *nitric acid* of the atmosphere, a substance whose functions towards vegetation are closely related to those of ammonia.

§ 7.

OZONE.

When lightning strikes the earth or an object near its surface, a person in the vicinity at once perceives a peculiar, so-called "sulphureous" odor, which must belong to something developed in the atmosphere by electricity. The same smell may be noticed in a room in which an electrical machine has been for some time in vigorous action.

The substance which is thus produced is termed *ozone*, from a Greek word signifying to smell. It is a colorless gas, possessing most remarkable properties, and is of the highest importance in agricultural science, although our knowledge of it is still exceedingly imperfect.

Ozone is not known in a pure state free from other bodies; but hitherto has only been obtained mixed with several times its weight of air or oxygen.* It is entirely insoluble in water. It has, when breathed, an irritating action on the lungs, and excites coughing like chlorine gas. Small animals are shortly destroyed in an atmosphere charged with it. It is itself instantly destroyed by a heat considerably below that of redness.

The special character of ozone that is of interest in connection with questions of agriculture is its *oxidizing power*. Silver is a metal which totally refuses to combine with oxygen under ordinary circumstances, as shown by its maintaining its brilliancy without symptom of rust or tarnish when exposed to pure air at common or at greatly elevated temperatures. When a slip of moistened silver is placed in a vessel the air of which is charged with ozone, the metal after no long time becomes coated with a black crust, and at the same time the ozone disappears.

By the application of a gentle heat to the blackened silver, *ordinary oxygen gas*, having the properties already mentioned as belonging to this element, escapes, and the slip recovers its original silvery color. The black crust is in fact an *oxide of silver* (AgO,) which readily suffers decomposition by heat. In a similar manner iron, copper, lead, and other metals, are rapidly oxidized.

A variety of vegetable pigments, such as indigo, litmus, etc., are speedily *bleached* by ozone. This action, also, is simply one of oxidation.

Gorup-Besanez (*Ann. Ch. u. Ph.*, 110, 86; also, *Physiologische Chemie*) has examined the deportment of a number of organic bodies towards ozone. He finds that egg-albumin and casein of milk are rapidly altered by it, while flesh fibrin is unaffected.

Starch, the sugars, the organic acids, and fats, are, when pure, unaffected by ozone. In presence of (dissolved in) *alkalies*, however, they are oxidized with more or less rapidity. It is remarkable that oxidation by ozone takes place only in the *presence of water*. Dry substances are unaffected by it.

The peculiar deportment towards ozone of certain volatile oils will be presently noticed.

---

* Babo and Claus (*Ann. Ch. u. Ph.*, 2d Sup., p. 304) prepared a mixture of oxygen and ozone containing nearly 6 per cent of the latter.

ATMOSPHERIC AIR AS THE FOOD OF PLANTS. 65

Tests for Ozone.—Certain phenomena of oxidation that are attended with changes of color serve for the recognition of ozone.

We have already seen (II. C. G., p. 64) that starch, when brought in contact with iodine, at once assumes a deep blue or purple color. When the compound of iodine with potassium, known as iodide of potassium, is acted on by ozone, its potassium is at once oxidized (to potash,) and the iodine is set free. If now paper be impregnated with a mixture of starch-paste and solution of iodide of potassium,* we have a test of the presence of ozone, at once most characteristic and delicate.

Such paper, moistened and placed in ozonous† air, is speedily turned blue by the action of the liberated iodine upon the starch. By the use of this test the presence and abundance of ozone in the atmosphere has been measured.

Ozone is Active Oxygen.—That ozone is nothing more or less than oxygen in a peculiar, active condition, is shown by the following experiment. When perfectly pure and dry oxygen is enclosed in a glass tube containing moist metallic silver in a state of fine division, it is possible by long-continued transmission of electrical discharges to cause the gaseous oxygen entirely to disappear. On heating the silver, which has become blackened (oxidized) by the process, the original quantity of oxygen is recovered in its ordinary state. The oxygen is thus converted under the influence of electricity into ozone, which unites with the silver and disappears in the solid combination.

The independent experiments of Andrews, Babo, and Soret, demonstrate that ozone has a greater density than oxygen, since the latter diminishes in volume when electrized. Ozone is therefore *condensed oxygen*,‡ i. e., its molecule contains more atoms than the molecule of ordinary oxygen gas.

---

* Mix 10 parts of starch with 200 parts of cold water and 1 part of recently fused iodide of potassium, by rubbing them together in a mortar; then heat to boiling, and strain through linen. Smear pure filter paper with this paste, and dry. The paper should be perfectly white, and must be preserved in a well-stoppered bottle.

† I. e., charged with ozone.

‡ Recent observations by Babo and Claus, and by Soret, show that the density of ozone is *one and a half times greater* than that of oxygen.

Allotropism.—This occurrence of an element in two or even more forms is not without other illustrations, and is termed Allotropism. Phosphorus occurs in two conditions, viz., red phosphorus, which crystallizes in rhombohedrons, and like ordinary oxygen is comparatively inactive in its affinities; and colorless phosphorus, which crystallizes in octahedrons, and, like ozone, has vigorous tendencies to unite with other bodies. Carbon is also found in three allotropic forms, viz., diamond, plumbago, and charcoal, which differ exceedingly in their chemical and physical characters.

Ozone Formed by Chemical Action.—Not only is ozone produced by electrical disturbance, but it has likewise been shown to originate from chemical action; and, in fact, from the very kind of action which it itself so vigorously manifests, viz., oxidation.

When a clean stick of colorless phosphorus is placed at the bottom of a large glass vessel, and is half covered with tepid water, there immediately appear white vapors, which shortly fill the apparatus. In a little time the peculiar odor of ozone is evident, and the air of the vessel gives, with iodide-of-potassium-starch paper, the blue color which indicates ozone. In this experiment ordinary oxygen, in the act of uniting with phosphorus, is partially converted into its active modification; and although the larger share of the ozone formed is probably destroyed by uniting with phosphorus, a portion escapes combination and is recognizable in the surrounding air.

The ozone thus developed is mingled with other bodies, (phosphorous acid, etc.,) which cause the white cloud. The quantity of ozone that appears in this experiment, though very small,—under the most favorable circumstances but $1/_{1300}$ of the weight of the air,—is still sufficient to exhibit all the reactions that have been described.

Schönbein has shown that various organic bodies which are susceptible of oxidation, viz., citric and tartaric acids, when dissolved in water and agitated with air in the sunlight for half an hour, acquire the reactions of ozone. Ether and alcohol, kept in partially filled bottles, also become capable of producing oxidizing effects. Many of the vegetable oils, as oil of turpentine, oil of lemon, oil of cinnamon, linseed oil, etc., possess the property of ozonizing oxygen, or at least acquire oxidizing properties when exposed to the air. Hence the bleaching and corrosion of the cork of a partially filled turpentine bottle.

It is a highly probable hypothesis that ozone may be formed in many or even all cases of slow oxidation, and that although the chief part of the ozone thus developed must unite at once with the oxidable substance, a portion of it may diffuse into the atmosphere and escape immediate combination.

Ozone is likewise produced in a variety of chemical reactions, whereby oxygen is liberated from combination at ordinary temperatures. When water is evolved by galvanic electricity into free oxygen and free hydrogen, the former is accompanied with a small proportion of ozone. The same is true in the electrolysis of carbonic acid. So, too, when permanganate of potash, binoxide of barium, or chromic acid, is mixed with strong sulphuric acid, oxygen gas is disengaged which contains an admixture of ozone.*

Is Ozone Produced by Vegetation?—It is an interesting question whether the oxygen so freely exhaled from the foliage of plants under the influence of sunlight is accompanied by ozone. Various experimenters have occupied

---

* It appears probable that ozone is developed in all cases of rapid oxidation at high temperatures. This has been long suspected, and Meissner obtained strong indirect evidence of the fact. Since the above was written, Pincus has announced that ozone is produced when hydrogen burns in the air, or in pure oxygen gas. The quantity of ozone thus developed is sufficient to be recognized by the odor. To observe this fact, a jet of hydrogen should issue from a fine orifice and burn with a small flame, not exceeding ⅜ inch in length. A clean, dry, and cold beaker glass is held over the flame for a few seconds, when its contents will smell as decidedly of ozone as the interior of a Leyden jar that has just been discharged. (Vs. St., IX, p. 473.) Pincus has also noticed the ozone odor in similar experiments with alcohol and oil (Argand) lamps, and with stearine candles.

Doubtless, therefore, we are justified in making the generalization that in all cases of oxidation ozone is formed, and in many instances a portion of it diffuses into the atmosphere and escapes immediate combination.

themselves with this subject. The most recent investigations of Daubeny, (*Journal Chem. Soc.*, 1867, pp. 1–28,) lead to the conclusion that ozone is exhaled by plants, a conclusion previously adopted by Scoutetten, Poey, De Luca, and Kosmann, from less satisfactory data. Daubeny found that air deprived of ozone by streaming through a solution of iodide of potassium, then made to pass the foliage of a plant confined in a glass bell and exposed to sunlight, acquired the power of blueing iodide-of-potassium-starch-paper, even when the latter was shielded from the light.* Cloëz, however, obtained the contrary results in a series of experiments made by him in 1855, (*Ann. de Chimie et de Phys.*, L, 326,) in which the oxygen, exhaled both from aquatic and land plants, contained in a large glass vessel, came into contact with iodide-of-potassium-starch-paper, situated in a narrow and blackened glass tube. Lawes, Gilbert, and Pugh, in their researches on the sources of the nitrogen of vegetation, (*Phil. Trans.*, 1861) examined the oxygen evolved from vegetable matter under the influence of strong light, without finding evidence of ozone. It is not impossible that ozone was really produced in the circumstances of Cloëz's experiments, but spent itself in some oxidizing action before it reached the test-paper. In Daubeny's experiments, however, the more rapid stream of air might have carried along over the test-paper enough ozone to give evidence of its presence. Although the question can hardly be considered settled, the evidence leads to the belief that vegetation itself is a source of ozone, and that this substance is exhaled, together with ordinary oxygen, from the foliage, when acted on by sunlight.

Ozone in the Atmosphere.—Atmospheric electricity, slow oxidation, and combustion, are obvious means of impregnating the atmosphere more or less with ozone. If

---

* Light alone blues this paper after a time in absence of ozone.

the oxygen exhaled by plants contains ozone, this substance must be perpetually formed in the atmosphere over a large share of the earth's surface.

The quantity present in the atmosphere at any one time must be very small, since, from its strong tendency to unite with and oxidize other substances, it shortly disappears, and under most circumstances cannot manifest its peculiar properties, except as it is continually reproduced. The ozone present in any part of the atmosphere at any given moment is then, not what has been formed, but what remains after oxidable matters have been oxidized. We find, accordingly, that atmospheric ozone is most abundant in winter; since then there not only occurs the greatest amount of electrical excitement * in the atmosphere, which produces ozone, but the earth is covered with snow, and thus the oxidable matters of its surface are prevented from consuming the active oxygen.

In the atmosphere of crowded cities, in the vicinity of manure heaps, and wherever considerable quantities of organic matters pervade the air, as revealed by their odor, there we find little or no ozone. There, however, it may actually be produced in the largest quantity, though from the excess of matters which at once combine with it, it cannot become manifest.

That the atmosphere ordinarily cannot contain more than the minutest quantities of ozone, is evident, if we accept the statement (of Schönbein?) that it communicates its odor distinctly to a million times its weight of air. The attempts to estimate the ozone of the atmosphere give varying results, but indicate a proportion far less than sufficient to be recognized by the odor, viz., not more than 1 part of ozone in 13 to 65 million of air. (Zwenger, Pless, and Pierre.)

These figures convey no just idea of the quantities of

---
* The amount of electrical disturbance is not measured by the number and violence of thunder-storms: these only indicate its intensity.

ozone actually produced in the atmosphere and consumed in it, or at the surface of the soil. We have as yet indeed no satisfactory means of information on this point, but may safely conclude from the foregoing considerations that ozone performs an important part in the economy of nature.

Relations of Ozone to Vegetable Nutrition.—Of the direct influence of atmospheric ozone on plants, nothing is certainly known. Theoretically it should be consumed by them in various processes of oxidation, and would have ultimately the same effects that are produced by ordinary oxygen.

Indirectly, ozone is of great significance in our theory of vegetable nutrition, inasmuch as it is the cause of chemical changes which are of the highest importance in maintaining the life of plants. This fact will appear in the section on Nitric Acid, which follows.

§ 8.

COMPOUNDS OF NITROGEN AND OXYGEN IN THE ATMOSPHERE.

Nitric Acid, NO_3H.—Under the more common name *Aqua fortis* (strong water) this highly important substance is to be found in every apothecary shop. It is, when pure, a colorless, usually a yellow liquid, whose most obvious properties are its sour, burning taste, and power of dissolving, or acting upon, many metals and other bodies.

When pure, it is a half heavier than its own bulk of water, and emits pungent, suffocating vapors or fumes; in this state it is rarely seen, being in general mixed or diluted with more or less water; when very dilute, it evolves no fumes, and is even pleasant to the taste.

It has the properties of an acid in the most eminent degree; vegetable blue colors are reddened by it, and it unites with great avidity to all basic bodies, forming a long list of nitrates.

It is volatile, and evaporates on exposure to air, though not so rapidly as water.

Nitric acid has a strong affinity for water; hence its vapors, when they escape into moist air, condense the moisture, making therewith a visible cloud or fume. For the same reason the commercial acid is always more or less dilute, it being difficult or costly to remove the water entirely.

Nitric acid, as it occurs in commerce, is made by heating together sulphuric acid and nitrate of soda, when nitric acid distils off, and sulphate of soda remains behind.

Nitrate of soda.	Sulphuric acid.	Bisulphate of soda.	Nitric acid.
$NO_3 Na$ +	$H_2 S O_4$ =	$HNa SO_4$ +	$NO_3 H$

Nitrate of soda is formed in nature, and exists in immense accumulations in the southern part of Peru, (see p. 252.)

Anhydrous Nitric Acid, N_2O_5, is what is commonly understood as existing in combination with bases in the nitrates. It is a crystallized body, but is not an acid until it unites with the elements of water.

Nitrate of Ammonia, NH_3 NO_3H, or $NH_4 NO_3$, may be easily prepared by adding to nitric acid, ammonia in slight excess, and evaporating the solution. The salt readily crystallizes in long, flexible needles, or as a fibrous mass. It gathers moisture from the air, and dissolves in about half its weight of water.

If nitrate of ammonia be mixed with potash, soda, or lime, or with the carbonates of these bases, an exchange of acids and bases takes place, the result of which is nitrate of potash, soda, or lime, on the one hand, and free ammonia or carbonate of ammonia on the other.

Nitrous Oxide, N_2O.—When nitrate of ammonia is heated, it melts, and gradually decomposes into water and nitrous oxide, a "laughing gas," as represented by the equation:—

$$NH_4 NO_3 = 2 H_2O + N_2O$$

Nitric acid and the nitrates act as powerful oxidizing agents, i. e., they readily yield up a portion or all their oxygen to substances having strong affinities for this element. If, for example, charcoal be warmed with strong nitric acid, it is rapidly acted upon and converted into carbonic acid. If thrown into melted nitrate of soda or saltpeter, it takes fire, and is violently burned to carbonic acid. Similarly, sulphur, phosphorus, and most of the metals, may be oxidized by this acid.

When nitric acid oxidizes other substances, it itself loses oxygen and suffers reduction to compounds of nitrogen, containing less oxygen. Some of these compounds require notice.

Nitric Oxide, NO.—When nitric acid somewhat diluted with water acts upon metallic copper, a gas is evolved, which, after washing with water, is colorless and permanent. It is nitric oxide. By exposure to air it unites with oxygen, and forms red, suffocating fumes of nitric peroxide, or, if the oxygen be not in excess, nitrous acid is formed.

Nitric Peroxide, (hyponitric acid,) NO_2, appears as a dark yellowish-red gas when strong nitric acid is poured upon copper or tin exposed to the air. It is procured in a state of purity by strongly heating nitrate of lead: by a cold approaching zero of Fahrenheit's thermometer, it may be condensed to a yellow liquid or even solid.

Nitrous Acid, (anhydrous,) N_2O_3, is produced when nitric peroxide is mixed with water at a low temperature, nitric acid being formed at the same time.

Nitric peroxide.   Water.   Nitric acid.   Nitrous acid, anhydrous.

$$4 NO_2 + H_2O = 2 NHO_3 + N_2O_3$$

It may be procured as a blue liquid, which boils at the freezing point of water.

When nitric peroxide is put in contact with solutions of an alkali, there results a mixture of nitrate and nitrite of the alkali.

Nitric   Hydrate of   Nitrate of   Nitrite of   Water.
peroxide.   potash.   potash.   potash.

$$2 NO_2 + 2 HKO = NKO_3 + NKO_2 + H_2O$$

Nitrite of Ammonia, $NH_4 NO_2$ is known to the chemist as a white crystalline solid, very soluble in water. When its concentrated aqueous solution is gently heated, the salt is gradually resolved into water and nitrogen gas. This decomposition is represented by the following equation:

$$NH_4 NO_2 = 2 H_2O + 2 N$$

This decomposition is, however, not complete. A portion of ammonia escapes in the vapors, and nitrous acid accumulates in the residual liquid. (Pettenkofer.) Addition of a strong acid facilitates decomposition; an alkali retards it. When a dilute solution, 1 : 500, is boiled, but a small portion of the salt is decomposed, and a part of it is found in the distillate. Very dilute solutions, 1 : 100,000, may be boiled without suffering any alteration whatever. (Schöyen.)

Schönbein and others have (erroneously?) supposed that nitrite of ammonia is generated by the direct union of nitrogen and water. Nitrite of ammonia may exist in the atmosphere in minute quantity.

Nitrites of potash and soda may be procured by strongly heating the corresponding nitrates, whereby oxygen gas is expelled.

The Mutual Convertibility of Nitrates and Nitrites is illustrated by various statements already made. There are, in fact, numerous substances which reduce nitrates to nitrites. According to Schönbein, (*Jour. Prakt. Ch.*, 84, 207,) this reducing effect is exercised by the albuminoids, by starch, glucose, and milk-sugar, but not by cane-sugar.

It is also manifested by many metals, as zinc, iron, and lead, and by any mixture evolving hydrogen, as, for example, putrefying organic matter. On the other hand, ozone instantly oxidizes nitrites to nitrates.

Reduction of Nitrates and Nitrites to Ammonia. — Some of the substances which convert nitrates into nitrites may also by their prolonged action transform the latter into ammonia. When small fragments of zinc and iron mixed together are drenched with warm solution of caustic potash, hydrogen is copiously disengaged. If a nitrate be added to the mixture, it is at once reduced, and ammonia escapes. If to a mixture of zinc or iron and dilute chlorhydric acid, such as is employed in preparing hydrogen gas, nitric acid, or any nitrate or nitrite be added, the evolution of hydrogen ceases, or is checked, and ammonia is formed in the solution, whence it can be expelled by lime or potash.

Nitric acid.   Hydrogen.   Ammonia.   Water.
$$NO_3H \;+\; 8H \;=\; NH_3 \;+\; 3H_2O$$

The appearance of nitrous acid in this process is an intermediate step of the reduction.

Further Reduction of Nitric and Nitrous Acids. — Under certain conditions nitric acid and nitrous acid are still further deoxidized. Nesbit, who first employed the reduction of nitric acid to ammonia by means of zinc and dilute chlorhydric acid as a means of determining the quantity of the former, mentions (*Quart. Jour. Chem. Soc.*, 1847, p. 283,) that when the temperature of the liquid is allowed to rise somewhat, *nitric oxide* gas, NO, escapes.

From weak nitric acid, zinc causes the evolution of nitrous oxide gas, N_2O.

As already mentioned, nitrate of ammonia, when heated to fusion, evolves nitrous oxide, N_2O. Emmet showed that by immersing a strip of zinc in the melted salt, nearly pure nitrogen gas is set free.

When nitric acid is heated with lean flesh (fibrin), nitric oxide and nitrogen gases both appear. It is thus seen that by successive steps of deoxidation nitric acid may be gradually reduced to nitrous acid, ammonia, nitric oxide, nitrous oxide, and finally to nitrogen.

Tests for Nitric and Nitrous Acids.—The fact that these substances often occur in extremely minute quantities renders it needful to employ very delicate tests for their recognition.

Price's Test.—Free nitrous acid decomposes iodide of potassium in the same manner as ozone, and hence gives a blue color, with a mixture of this salt and starch-paste. Any nitrite produces the same effect if to the mixture dilute sulphuric acid be added to liberate the nitrous acid. Pure nitric acid, if moderately dilute, and dilute solutions of nitrates mixed with dilute sulphuric acid, are without immediate effect upon iodide-of-potassium-starch-paste. If the solution of a nitrate be mingled with dilute sulphuric acid, and agitated for some time with zinc filings, reduction to nitrite occurs, and then addition of the starch-paste, etc., gives the blue coloration. According to Schönbein, this test, first proposed by Price, will detect nitrous acid when mixed with one-hundred-thousand times its weight of water. It is of course only applicable in the absence of other oxidizing agents.

Green Vitriol Test.—A very characteristic test for nitric and nitrous acids, and a delicate one, though less sensitive than that just described, is furnished by common green vitriol, or protosulphate of iron. Nitric oxide, the red gas which is evolved from nitric acid or nitrates by mixing them with excess of *strong sulphuric acid*, and from nitrous acid or nitrites by addition of *dilute sulphuric acid*, gives with green vitriol a peculiar blackish-brown coloration. To test for minute quantities of nitrous acid, mix the solution with dilute sulphuric acid and cautiously pour this liquid upon an equal bulk of cold saturated solution of green vitriol, so that the former liquid floats upon the latter without mingling much with it. On standing, the coloration will be perceived where the two liquids are in contact.

Nitric acid is tested as follows: Mix the solution of nitrate with an equal volume of concentrated sulphuric acid; let the mixture cool, and pour upon it the solution of green vitriol. The coloration will appear between the two liquids.

Formation of Nitrogen Compounds in the Atmosphere.

—*a*. From free nitrogen, by electrical ozone. Schönbein and Meissner have demonstrated that a discharge of electricity through air in its ordinary state of dryness causes oxygen and nitrogen to unite, with the formation of nitric peroxide, NO_2. Meissner has proved that not the electricity directly, but the *ozone* developed by it, accomplishes this oxidation. It has long been known that nitric peroxide decomposes with water, yielding nitric and nitrous acids thus:

$$2 NO_2 + H_2O = NO_3H + NO_2H.$$

It is further known that nitrous acid, both in the free state and in combination, is instantly oxidized to nitric acid by contact with ozone.

Thus is explained the ancient observation, first made by Cavendish in 1784, that when electrical sparks are transmitted through moist air, confined over solution of potash, nitrate of potash is formed. (For information regarding this salt, see p. 252.)

Until recently, it has been supposed that nitric acid is present in only those rains which accompany thunder-storms.

It appears, however, from the analyses of both Way and Boussingault, that visible or audible electric discharges do not perceptibly influence the proportion of nitric acid in the air; the rains accompanying thunder-storms not being always nor usually richer in this substance than others.

Von Babo and Meissner have demonstrated that *silent electrical discharges* develop more ozone than flashes of lightning. Meissner has shown that the electric spark causes the copious formation of nitric peroxide in its immediate path by virtue of the heat it excites, which increases the energy of the ozone simultaneously produced, and causes it to expend itself at once in the oxidation of nitrogen. Boussingault informs us that in some of the tropical regions of South America audible electrical discharges are continually taking place throughout the whole year. In our latitudes electrical disturbance is perpetually occurring, but equalizes itself mostly by silent discharge. The ozone thus noiselessly developed, though operating at a lower temperature, and therefore more slowly than that which is produced by lightning, must really oxidize much more nitrogen to nitric acid than the latter, because its action never ceases.

Formation of Nitrogen Compounds in the Atmosphere.
—*b*. From free nitrogen (by ozone?) in the processes of combustion and slow oxidation.

At high temperatures.—Saussure first observed (*Ann. de Chimie*, lxxi, 282), that in the burning of a mixture of oxygen and hydrogen gases in the air, the resulting water contains ammonia. He had previously noticed that nitric acid and nitrous acid are formed in the same process.

Kolbe (*Ann. Chem. u. Pharm.*, cxix, 176) found that when a jet of burning hydrogen was passed into the neck of an open bottle containing oxygen, reddish-yellow vapors of nitrous acid or nitric peroxide were copiously produced on atmospheric air becoming mingled with the burning gases.

Bence Jones (*Phil. Trans.*, 1851, ii, 399) discovered nitric (nitrous?) acid in the water resulting from the burning of alcohol, hydrogen, coal, wax, and purified coal-gas.

By the use of the iodide-of-potassium-starch test (Price's test), Boettger (*Jour. für Prakt. Chem.*, lxxxv, 396) and Schönbein (ibid., lxxxiv, 215) have more recently confirmed the result of Jones, but because they could detect neither free acid nor free alkali by the ordinary test-papers, they concluded that nitrous acid and ammonia are simultaneously formed, that, in fact, *nitrite of ammonia* is generated in all cases of rapid combustion.

Meissner (*Untersuchungen über den Sauerstoff*, 1863, p. 283) was unable to satisfy himself that either nitrous acid or ammonia is generated in combustion.

Finally, Zabelin (*Ann. Chem. u. Ph.*, cxxx, 54) in a series of careful experiments, found that when alcohol, illuminating gas, and hydrogen, burn in the air, nitrous acid and ammonia are very frequently, but not always, formed.

When the combustion is so perfect that the resulting water is colorless and pure, only nitrous acid is formed; when, on the other hand, a trace of organic matters escapes oxidation, less or no nitrous acid, but in its place ammonia, appears in the water, and this under circumstances that preclude its absorption from the atmosphere.

Zabelin gives no proof that the combustibles he employed were absolutely free from compounds of nitrogen, but otherwise, his experiments are not open to criticism.

Meissner's observations were indeed made under somewhat different conditions; but his negative results were not improbably arrived at simply because he employed a much less delicate test for nitrous acid than was used by Schönbein, Boettger, Jones, and Zabelin.*

We must conclude, then, that nitrous acid and ammonia are usually formed from atmospheric nitrogen during rapid combustion of hydrogen and compounds of hydrogen and carbon. The quantity of these bodies thus generated is, however, in general so extremely small as to require the most sensitive reagents for their detection.

At low temperatures.—Schönbein was the first to observe that nitric acid may be formed at moderately elevated or even ordinary temperatures. He obtained several grams of nitrate of potash by adding carbonate of potash to the liquid resulting from the slow oxidation of phosphorus in the preparation of ozone.

More recently he believed to have discovered that nitrogen compounds are formed by the simple evaporation of water. He heated a vessel (which was indifferently of

---

* Meissner rejected Price's test in the belief that it cannot serve to distinguish nitrous acid from peroxide of hydrogen, H_2O_2. He therefore made the liquid to be examined alkaline with a slight excess of potash, concentrated to small bulk and tested with dilute sulphuric acid and protosulphate of iron. (*Unters. ü. d. Sauerstoff*, p. 233). Schönbein had found that iodide of potassium is decomposed after a little time by *concentrated* solutions of peroxide of hydrogen, but is unaffected by this body when dilute. (*Jour. für prakt. Chem.*, lxxxvi. p. 90). Zabelin agrees with Schönbein that Price's test is decisive between peroxide of hydrogen and nitrous acid. (*Ann. Chem. u. Ph.*, cxxx, p. 58.)

glass, porcelain, silver, etc.,) so that water would evaporate rapidly from its surface. The purest water was then dropped into the warm dish in small quantities at a time, each portion being allowed to evaporate away before the next was added. Over the vapor thus generated was held the mouth of a cold bottle until a portion of the vapor was condensed in the latter.

The water thus collected gave the reactions for nitrous acid and ammonia, sometimes quite intensely, again faintly, and sometimes not at all.

By simply exposing a piece of filter-paper for a sufficient time to the vapors arising from pure water heated to boiling, and pouring a few drops of acidified iodide-of-potassium-starch-paste upon it, the reaction of nitrous acid was obtained. When paper which had been impregnated with dilute solution of pure potash was hung in the vapors that arose from water heated in an open dish to 100° F., it shortly acquired so much nitrite of potash as to react with the above named test.

Lastly, nitrous acid and ammonia appeared when a sheet of filter-paper, or a piece of linen cloth, which had been moistened with the purest water, was allowed to dry at ordinary temperatures, in the open air or in a closed vessel. (*Jour. für Prakt. Chem.*, lxvi, 131.) These experiments of Schönbein are open to criticism, and do not furnish perfectly satisfactory evidence that nitrous acid and ammonia are generated under the circumstances mentioned. Bohlig has objected that these bodies might be gathered from the atmosphere, where they certainly exist, though in extremely minute quantity.

Zabelin, in the paper before referred to (*Ann. Ch. Ph.* cxxx, p. 76), communicates some experimental results which, in the writer's opinion, serve to clear up the matter satisfactorily.

Zabelin ascertained in the first place that the atmospheric air contained too little ammonia to influence Nessler's test,* which is of extreme delicacy, and which he constantly employed in his investigations.

Zabelin operated in closed vessels. The apparatus he used consisted of two glass flasks, a larger and a smaller one, which were closed by corks and fitted with glass tubes, so that a stream of air entering the larger vessel should bubble through water covering its bottom, and thence passing into the smaller flask should stream through Nessler's test. Next, he found that no ammonia and (by Price's test) but doubtful traces of nitrous acid could be detected in the purest water when distilled alone in this apparatus.

Zabelin likewise showed that cellulose (clippings of filter-paper or shreds of linen) yielded no ammonia to Nessler's test when heated in a current of air at temperatures of 120° to 160° F.

Lastly, he found that when cellulose and pure water together were exposed to a current of air at the temperatures just named, ammonia was at once indicated by Nessler's test. Nitrous acid, however, could be detected, if at all, in the minutest traces only.

Views of Schönbein.—The reader should observe that Boettger and Schönbein, finding in the first instance by the exceedingly sensitive test with iodide of potassium and starch-paste, that nitrous acid was formed, when hydrogen burned in the air, while the water thus generated was neutral in its reaction with the *vastly less sensitive* litmus test-paper, concluded that the nitrous acid was united with some base in the form of a neutral salt. Afterward, the detection of ammonia appeared to demonstrate the formation of *nitrite of ammonia.*

We have already seen that nitrite of ammonia, by exposure to a moderate heat, is resolved into nitrogen and water. Schönbein assumed that under the conditions of

---

* See p. 54.

his experiments nitrogen and water combine to form nitrite of ammonia.

$$2 N + 2 H_2O = NH_3, NO_2H$$

This theory, supported by the authority of so distinguished a philosopher, has been almost universally credited.* It has, however, little to warrant it, even in the way of probability. If traces of nitrite of ammonia can be produced by the immediate combination of these exceptionally abundant and universally diffused bodies at common temperatures, or at the boiling point of water, or lastly in close proximity to the flames of burning gases, then it is simply inconceivable that a good share of the atmosphere should not speedily dissolve in the ocean, for the conditions of Schönbein's experiments prevail at all times and at all places, so far as these substances are concerned.

The discovery of Zabelin that ammonia and nitrous acid do not always appear in equivalent quantities or even simultaneously, while difficult to reconcile with Schönbein's theory, in no wise conflicts with any of his facts. A quantity of free nitrous acid that admits of recognition by help of Price's test would not necessarily have any effect on litmus or other test for free acids. There remains, then, no necessity of assuming the generation of nitrite of ammonia, and the fact of the separate appearance of the elements of this salt demands another explanation.

The Author's Opinion.—The writer is not able, perhaps, to offer a fully satisfactory explanation of the facts above adduced. He submits, however, some speculations which appear to him entirely warranted by the present aspects of the case, in the hope that some one with the time at

---

* Zabelin was inclined to believe that his failure to detect nitrous acid in some of his experiments where organic matters intervened, was due to a power possessed by these organic matters to mask or impair the delicacy of Price's test, as first noticed by Pettenkofer and since demonstrated by Schönbein in case of urine.

command for experimental study, will establish or disprove them by suitable investigations.

He believes, from the existing evidence, that free nitrogen can, in no case, unite directly with water, but in the conditions of all the foregoing experiments, it enters combination by the action of *ozone*, as Schönbein formerly maintained and was the first to suggest.

We have already recounted the evidence that goes to show the formation of ozone in all cases of oxidation, both at high and low temperatures, p. 67.

In Zabelin's experiments we may suppose that ozone was formed by the oxidation of the cellulose (linen and paper) he employed. In Schönbein's experiments, where paper or linen was not employed, the dust of the air may have supplied the organic matters.

The first result of the oxidation of nitrogen is nitrous acid alone (at least Schönbein and Bohlig detected no nitric acid), when the combustion is complete, as in case of hydrogen, or when organic matters are excluded from the experiment. Nitric acid is a product of the subsequent oxidation of nitrous acid. When organic matters exist in the product of combustion, as when alcohol burns in a heated apparatus yielding water having a yellowish color, it is probable that nitrous acid is formed, but is afterward *reduced* to ammonia, as has been already explained, p. 74.

Zabelin, in the article before cited, refers to Schönbein as authority for the fact that various organic bodies, viz., all the vegetable and animal albuminoids, gelatine, and most of the carbohydrates, especially starch, glucose, and milk-sugar, reduce nitrites to *ammonia*, and ultimately to nitrogen; and although we have not been able to find such a statement in those of Schönbein's papers to which we have had access, it is entirely credible and in accordance with numerous analogies.

If, as thus appears extremely probable, ozone is developed in all cases of oxidation, both rapid and slow, then every flame and fire, every decaying plant and animal, the organic matters that exhale from the skin and lungs of living animals, or from the foliage and flowers of plants, especially, perhaps, the volatile oils of cone-bearing trees, are, indirectly, means of converting a portion of free nitrogen into nitrous and nitric acids, or ammonia.

These topics will be recurred to in our discussion of Nitrification in the Soil, p. 254.

Formation of Nitrogen Compounds in the Atmosphere.
—*c.* From free nitrogen by ozone accompanying the oxygen exhaled from green foliage in sunlight.

The evidence upon the question of the emission of ozone by plants, or of its formation in the vicinity of foliage, has been briefly presented on page 68. The present state of investigation does not permit us to pronounce definitely upon this point. There are, however, some facts of agriculture which, perhaps, find their best explanation by assuming this evolution of ozone.

It has long been known that certain crops are especially aided in their growth by nitrogenous fertilizers, while others are comparatively indifferent to them. Thus the cereal grains and grasses are most frequently benefited by applications of nitrate of soda, Peruvian guano, dung of animals, fish, flesh and blood manures, or other matters rich in nitrogen. On the other hand, clover and turnips flourish best, as a rule, when treated with phosphates and alkaline substances, and are not manured with animal fertilizers so economically as the cereals. It has, in fact, become a rule of practice in some of the best farming districts of England, where systematic rotation of crops is followed, to apply nitrogenous manures to the cereals and phosphates to turnips. Again, it is a fact, that whereas nitrogenous manures are often necessary to produce a good wheat crop, in which, at 30 bu. of grain and 2,600 lbs. of straw, there is contained 45 lbs. of nitrogen; a crop of clover may be produced without nitrogenous manure, in which would be taken from the field twice or thrice the above amount of nitrogen, although the period of growth of the two crops is about the same. Ulbricht found in his investigation of the clover plant (*Vs. St.*, IV., p. 27) that the soil appears to have but little influence on the content of nitrogen of clover, or of its individual organs. These facts admit of another expression, viz.: Clover, though containing two or three times more nitrogen, and requiring correspondingly larger supplies of nitrates and ammonia than wheat, *is able to supply itself* much more easily than the latter crop. In parts of the Genesee wheat region, it is the custom to alternate clover with wheat, because the decay of the clover stubble and roots admirably prepares the ground for the last-named crop. The same preparation might be had by the more expensive process of dressing with a highly nitrogenous manure, and it is scarcely to be doubted that it is the *nitrogen* gathered by the clover which insures the wheat crop that follows. It thus appears that the plant itself causes the formation in its neighborhood of assimilable compounds of nitrogen, and that some plants excel others in their power of accomplishing this important result.

On the supposition that ozone is emitted by plants, it is plain that those crops which produce the largest mass of foliage develop it most abundantly. By the action of this ozone, the nitrogen that bathes the leaves is converted into nitric acid, which, in its turn, is absorbed by the plant. The foliage of clover, cut green, and of root crops, maintains its activity until the time the crop is gathered; the supply of nitrates thus keeps pace with the wants of the plant. In case of grain crops, the functions of the foliage decline as the seed begins to develop, and the plant's means of providing itself with assimilable nitrogen fail, although the need for it still exists. Furthermore, the clover cut for hay, leaves behind much more roots and stubble per acre than grain crops, and the clover stubble is twice as rich in nitrogen as the stubble of ripened grain. This is a result of the fact that the clover is cut when in active growth, while the grain is harvested after the roots, stems, and leaves, have been exhausted of their own juices to meet the demands of the seed.

Whatever may be the value of our explanations, the fact is not to be denied that the soil is enriched in nitrogen by the culture of large-leaved plants, which are harvested while in active growth, and leave a considerable proportion of roots, leaves, or stubble, on the field. On the other hand, the field is impoverished in nitrogen when grain crops are raised upon it.

Formation of Nitric Acid from Ammonia.—Ammonia (carbonate of ammonia) under the influence of ozone is converted into nitrate of ammonia, (Baumert, Houzeau). The reaction is such that one-half of the ammonia is oxidized to nitric acid, which unites with the residue and with water, as illustrated by the equation:

$$2 NH_3 + 4 O = NH_4, NO_3 + H_2O$$

In this manner, nitrate of ammonia may originate in the atmosphere, since, as already shown, ammonia and ozone are both present there.

Oxidation and Reduction in the Atmosphere.—The fact that ammonia and organic matters on the one hand, and ozone, nitrous and nitric acids on the other, are present, and, perhaps, constantly present in the air, involves at first thought a contradiction, for these two classes of substances are in a sense incompatible with each other. Organic matters, ammonia, and nitrous acid, are converted by ozone into nitric acid. On the contrary, certain organic matters reduce ozone to ordinary oxygen, or destroy it altogether, and reduce nitric and nitrous acids to ammonia, or, perhaps, to free nitrogen. The truth is that the substances named are being perpetually composed and decomposed in the atmosphere, and at the surface of the soil. Here, or at one moment, oxidation prevails; there, or at another moment, reduction preponderates. It is only as one or another of the results of this incessant action is withdrawn from the sphere of change, that we can give it permanence and identify it. The quantities we measure are but resultants of forces that oppose each other. The idea of rest or permanence is as foreign to the chemistry of the atmosphere as to its visible phenomena.

Nitric Acid in the Atmosphere.—The occurrence of nitric acid or nitrate of ammonia in the atmosphere has been abundantly demonstrated in late years (1854-6) by Cloez, Boussingault, De Luca, and Kletzinsky, who found that when large volumes of air are made to bubble through solutions of potash, or to stream over fragments of brick or pumice which have been soaked in potash or carbonate of potash, these absorbents gradually acquire a small amount of nitric acid. In the experiments of Cloez and De Luca, the air was first washed of its ammonia by contact with sulphuric acid. Their results prove, therefore, that the nitric acid was formed independently of ammonia, though it doubtless exists in the air in combination with this base.

Proportion of Nitric Acid in Rain-water, etc.—In atmospheric waters, nitric acid is found much more abundantly than in the air itself, for the reason that a small bulk of rain, etc., washes an immense volume of air.

Many observers, among the first, Liebig, have found nitrates in rain-water, especially in the rain of thunderstorms. The investigations of Boussingault, made in 1856-8, have amply confirmed Barral's observation that nitric acid (in combination) is almost invariably present in rain, dew, fog, hail, and snow. Boussingault, (*Agronomie, etc.*, II, 325) determined the quantity of nitric acid in 134 rains, 31 snows, 8 dews, and 7 fogs. In only 16 instances out of these 180 was the amount of nitric acid too small to detect. The greatest proportion of nitric acid found in rain occurred in a slow-falling morning shower, (9th October, 1857, at Liebfrauenberg), viz., 62 parts* in 10 million of water. In fog, on one occasion, (at Paris, 19th Dec., 1857,) 101 parts to 10 million of water were observed.

Knop found in rain-water, collected near Leipzig, in July, 1862, 56 parts; in rain that fell during a thunderstorm, 98 parts in 10 million of water.

Boussingault found in rain an average of 2 parts, in snow of 4 parts, of nitric acid to 10 million of water.

Mr. Way, whose determinations of ammonia in the atmospheric waters collected by Lawes and Gilbert, at Rothamstead, during the whole of the years 1855-6, have already been noticed, (p. 63,) likewise estimated the nitric acid in the same waters. He found the proportion of nitric acid to be, in 1855, 4 parts, in 1856 $4\frac{1}{2}$ parts, to 10 million of water.

Bretschneider found at Ida-Marienhütte, Prussia, for the year 1865-6 an average of $8\frac{1}{2}$ parts, for 1866-7 an average of $4\frac{1}{2}$ parts, of nitric acid in 10 million of water. At Regenwalde, Prussia, the average in 1865-6 was 25 parts, in 1866-7, 22 parts. At Proskau, the average in 1864-5 was 31 parts. At Kuschen, the average for 1864-5 was 6 parts; in 1865-6, 7; in 1866-7, 8 parts. At Dahme, in 1865-6, the average was 12 parts. At Insterburg, Pincus and Röllig obtained in 1864-5, an average of 12 parts; in 1865-6, an average of 16 parts of nitric acid in 10 million of water. The highest monthly average was 280 parts, at Lauersfort, July, 1864; and the lowest was nothing, April, 1865, at Ida-Marienhütte.

Quantity of Nitric Acid in Atmospheric Water.—The total quantity of nitric acid that could be collected in the rains, etc., at Rothamstead, amounted in 1855 to 2.98 lbs., and in 1856 to 2.80 lbs. per acre.

---

* In all the quantitative statements here and elsewhere, *anhydrous nitric acid*, N_2O_5, (O=16, formerly NO_5, O=8) is to be understood.

This quantity was very irregularly distributed among the months. In 1855 the smallest amount was collected in January, the largest in October, the latter being nearly 20 times as much as the former. In 1856 the largest quantity occurred in May, and the smallest in February, the former not quite six times as much as the latter.

The following table gives the results of Mr. Way entire. (*Jour. Roy. Ag. Soc. of Eng.*, XVII, pp. 144 and 620.)

AMOUNTS OF RAIN AND OF AMMONIA, NITRIC ACID, AND TOTAL NITROGEN therein, collected at Rothamstead, Eng., in the years 1855–6—calculated per acre, according to Messrs. Lawes, Gilbert, and Way.

	Quantity of Rain in Imperial Gallons, (1 gal. = 10 lbs. water.)		Ammonia in grains.		Nitric acid in grains.		Total Nitrogen in grains.	
	1855	1856	1855	1856	1855	1856	1855	1856
January	13.523	62.952	1244	5005	230	1561	1084	4526
February	22.473	30.586	2337	4175	944	544	2169	3579
March	52.484	22.722	4513	2108	1102	806	3995	1945
April	9.281	59.083	1141	8614	325	1063	1024	7369
May	52.575	106.474	4206	18313	1810	3024	3089	15863
June	41.295	43.253	5574	4870	3308	2046	5447	4540
July	157.713	33.561	9620	2869	2680	1191	8615	2670
August	59.622	59.850	4769	4214	3577	2125	4870	4021
September	34.875	47.477	3313	5972	732	1756	2917	5373
October	124.466	65.038	7592	3921	4480	2075	7414	3767
November	59.950	32.181	3021	2591	1007	1371	2749	2489
December	39.075	50.870	2438	4070	664	2035	2180	3352
Total	663.332 gall's.	616.051 gall's.	7.11 lbs.	9.53 lbs.	2.98 lbs.	2.80 lbs.	6.63 lbs.	8.31 lbs.

According to Pincus and Röllig, the atmospheric water brought down at Insterburg, in the year ending with March, 1865, 7.225 lbs. av. of nitric acid per English acre of surface.

The quantity of nitrogen that fell as ammonia was 3.628 lbs.; that collected in the form of nitric acid was 1.876 lbs. The total nitrogen of the atmospheric waters per acre, for the year, was 5.5 lbs. The rain-fall was 392.707 imperial gallons.

Bretschneider found in the atmospheric waters gathered at Ida-Marienhütte, in Silesia, during 12 months ending April 15th, 1866, 3¾ lbs. of nitric acid per acre of surface. In Bretschneider's investigation, the amount of nitrogen brought down per acre in the form of ammonia was 9.936 lbs.; that in the form of nitric acid was 0.974 lbs. The total nitrogen contained in the rain, etc., was accordingly 10.91, or, in round numbers, 11 lbs. avoirdupois. The rainfall amounted to 488.309 imperial gallons, (Wilda's *Centralblatt*, August, 1866.)

Relation of Nitric Acid to Ammonia in the Atmosphere.—The foregoing results demonstrate that there is in the aggregate an excess of ammonia over the amount required to form nitrate with the nitric acid. (In nitrate of ammonia ($NH_4\ NO_3$), the acid and base contain the same quantity of nitrogen.) We are hence justified in assuming that the acid in question commonly occurs as nitrate of ammonia * in the atmosphere.

At times, however, the nitric acid may preponderate. One instance is on record (*Journal de Pharmacie*, Apr., 1845) of the presence of free nitric acid in hail, which fell at Nismes, in June, 1842. This hail is said to have been perceptibly sour to the taste.

Cloez (*Compt. Rendus*, lii, 527) found traces of free nitric acid in air taken 3 feet above the ground, especially at the beginning and end of winter.

The same must have been true in the cases already given, in which exceptionally large quantities of nitric acid were found, in the examinations made by Boussingault and the Prussian chemists.

The nitrate of ammonia which exists in the atmosphere doubtless held there in a state of mechanical suspension. t is dissolved in the falling rains, and when once brought to the surface of the soil, cannot again find its way into the air by volatilization, as carbonate of ammonia does, but is permanently removed from the atmosphere, and

---

* In evaporating large quantities of rain-water to dryness, there are often found in the residue nitrates of lime and soda. In these cases the lime and soda come from dust suspended in the air.

until in some way chemically decomposed, belongs to the soil or to the rivers and seas.

Nitrous Acid in the Atmospheric Waters.—In most of the researches upon the quantity of nitric acid in the atmosphere and meteoric waters, *nitrous acid* has not been specially regarded. The tests which serve to detect nitric acid nearly all apply equally well to nitrous acid, and no discrimination has been made until recently. According to Schönbein and Bohlig, nitrates are sometimes absent from rain-water, but nitrites never. They occur, however, in but minute proportion. Pincus and Röllig observed but traces of nitrous acid in the waters gathered at Insterburg. Reichardt found no weighable quantity of nitrous acid in a sample of hail, the water from which contained in 10 million parts, 32 parts ammonia and $5\frac{1}{2}$ parts of nitric acid. It is evident, then, that nitrous acid, if produced to any extent in the atmosphere, does not remain as such, but is chiefly oxidized to nitric acid.

In any case our data are probably not incorrect in respect to the quantity of *nitrogen* existing in both the forms of nitrous and nitric acids, although the former compound has not been separately estimated. The methods employed for the estimation of nitric acid would, in general, include the nitrous acid, with the single error of bringing the latter into the reckoning as a part of the former.

Nitric Acid as Food of Plants.—A multitude of observations, both in the field and laboratory, demonstrate that nitrates greatly promote vegetable growth. The extensive use of nitrate of soda as a fertilizer, and the extraordinary fertility of the tropical regions of India, whose soil until lately furnished a large share of the nitrate of potash of commerce, attest the fact. Furthermore, in many cases, nitrates have been found abundantly in fertile soils of temperate climates.

Experiments in artificial soil and in water-culture show not only that nitrates supply nitrogen to plants, but demonstrate beyond doubt that *they alone are a sufficient source* of this element, and that no other compound is so well adapted as nitric acid to furnish crops with nitrogen.

Like ammonia-salts, the nitrates intensify the color, and increase, both absolutely and relatively, the quantity of nitrogen of the plant to which they are supplied. Their effect, when in excess, is also to favor the development of foliage at the expense of fruit.

ATMOSPHERIC AIR AS THE FOOD OF PLANTS. 91

The nitrates do not appear to be absorbed by the plant to any great extent, except through the medium of the soil, since they cannot exist in the state of vapor and are brought down to the earth's surface by atmospheric waters.

The full discussion of their nutritive effects must therefore be deferred until the soil comes under notice. See Division II, p. 271.

In § 10, p. 96, "Recapitulation of the Atmospheric Supplies of Food to Crops," the inadequacy of the atmospheric nitrates will be noticed.

§ 9.

OTHER INGREDIENTS OF THE ATMOSPHERE; viz., *Marsh Gas, Carbonic Oxide, Nitrous Oxide, Hydrochloric Acid, Sulphurous Acid, Sulphydric Acid, Organic Vapors, Suspended Solid Matters.*

There are several other gaseous bodies, some or all of which may occur in the atmosphere in very minute quantities, but whose relations to vegetation, in the present state of our knowledge, appear to be of no practical moment. Since, however, they have been the subjects of investigations or disquisition by agricultural chemists, they require to be briefly noticed.

Marsh Gas,[*] CH_4.—This substance is a colorless and nearly odorless gas, which is formed almost invariably when organic matters suffer decomposition in absence of oxygen. When a lump of coal or a billet of wood is strongly heated, portions of carbon and hydrogen unite to form this among several other substances. It is accordingly one of the ingredients of the gases whose combustion forms the flame of all fires and lamps. It is also produced in the decay of vegetable matters, especially when they are immersed in water, as happens in swamps and stagnant ponds, and it often bubbles in large quantities from the bottom of ditches, when the mud is stirred.

Pettenkofer and Voit have lately found that marsh gas is one of the gaseous products of the respiration or nutrition of animals.

It is combustible at high temperatures, and burns with a yellowish, faintly luminous flame, to water and carbonic acid. It causes no ill effects when breathed by animals if it be mixed with much air, though of itself it cannot support respiration.

---

[*] Known also to chemists under the names of Light Carburetted Hydrogen, Hydride of Methyl and Methane.

The mode of its origin at once suggests its presence in the atmosphere. Saussure observed that common air contains some gaseous compound or compounds of carbon, besides carbonic acid; and Boussingault found in 1834 that the air at Paris contained a very small quantity (from two to eight-millionths) of hydrogen in some form of combination besides water. These facts agree with the supposition that marsh gas is a normal though minute and variable ingredient of the atmosphere.

Relations of Marsh Gas to Vegetation.—Whether this gas is absorbed and assimilated by plants is a point on which we have at present no information. It might serve as a source both of carbon and hydrogen; but as these bodies are amply furnished by carbonic acid and water, and as it is by no means improbable that marsh gas itself is actually converted into these substances by ozone, the question of its assimilation is one of little importance, and remains to be investigated.

Schultz (Johnston's *Lectures on Ag. Chem.*, 2d Ed., 147) found on several occasions that the gas evolved from plants when exposed to the sunlight, instead of being pure oxygen, contained a combustible admixture, so that it exploded violently on contact with a lighted taper.

This observation shows either that the healthy plants evolved a large amount of marsh gas, which forms with oxygen an explosive mixture (the fire-damp of coal-mines), or, as is most probable, that the vegetable matter entered into decomposition from too long continuance of the experiment.

Boussingault has, however, recently found a minute proportion of marsh gas in the air exhaled from the leaves of plants that are exposed to sunlight *when submerged in water*. It does not appear when the leaves are surrounded by air, as the latest experiments of Boussingault, Cloez, and Corenwinder, agree in demonstrating.

Carbonic Oxide. CO, is a gas destitute of color and odor. It burns in contact with air, with a flame that has a fine blue color. The result of its combustion is carbonic acid, $CO + O = CO_2$.

This gas is extremely poisonous to animals. Air containing a few per cent of it is unfit for respiration, and produces headache, insensibility, and death.

Carbonic oxide may be obtained artificially by a variety of processes. If carbonic acid gas be made to stream slowly through a tube containing ignited charcoal, it is converted into carbonic oxide, $CO_2 + C = 2\,CO$.

Carbonic oxide is largely produced in all ordinary fires. The air which draws through a grate heaped with well-ignited coals, as it enters the bottom of the mass of fuel, loses a large portion of its oxygen, which there unites with carbon, forming carbonic acid. This gas is carried up into the heated coal, and there, where carbon is in excess, it takes up another proportion of this element, being converted into carbonic oxide. At the summit of the fire, where oxygen is abundant, the carbonic oxide burns again with its peculiar blue color, to carbonic acid, provided the heat be intense enough to inflame the gas, as is the case when the mass of fuel is thoroughly ignited. When, on the other hand, the fire is covered with cold fuel, carbonic oxide escapes copiously into the atmosphere.

When crystallized oxalic acid is heated with oil of vitriol, it yields water to the latter, and falls into a mixture of carbonic acid and carbonic oxide.

$$C_2 H_2 O_4, 2H_2O = CO_2 + CO + 3 H_2O.$$

Carbonic oxide may, perhaps, be formed in small quantity in the decay of organic matters; though Corenwinder (*Compt. Rend.*, LX, 102) failed to detect it in the rotting of manure.

Relations of Carbonic Oxide to Vegetation.—According to Saussure, while pea-plants languish and die when immersed in carbonic oxide, certain marsh plants (*Epilobium hirsutum*, *Lythrum salicaria*, and *Polygonum persicaria*) flourish as well in this gas as in common air. Saussure's experiments with these plants lasted six weeks. There occurred an absorption of the gas and an evolution of oxygen. It is thus to be inferred that carbonic oxide may be a source of carbon to aquatic plants.

Boussingault (*Compt. Rend.*, LXI, 493) was unable to detect any action of the foliage of land plants upon carbonic oxide, either when the gas was pure or mixed with air.

The carbonic oxide which Boussingault found in 1863 in air exhaled from submerged leaves, proves to have been produced in the analyses, (from pyrogallate of potash,) and was not emitted by the leaves themselves, as at first supposed, as both Cloez and Boussingault have shown.

Nitrous Oxide. N_2O.—This substance, the so-called *laughing gas*, is prepared from nitrate of ammonia by exposing that salt to a heat somewhat higher than is necessary to fuse it. The salt decomposes into nitrous oxide and water.

$$NH_4, NO_3 = N_2O + 2 H_2O.$$

The gas is readily soluble in water, and has a sweetish odor and taste. When breathed, it at first produces a peculiar exhilarating effect, which is followed by stupor and insensibility.

This gas has never been demonstrated to exist in the atmosphere. In fact, our methods of analysis are incompetent to detect it, when it is present in very minute quantity in a gaseous mixture. Knop is of the opinion that nitrous oxide may occur in the atmosphere, and has published an account of experiments (*Journal für Prakt. Chem.*, Vol. 59, p. 114) which, according to him, prove that it is absorbed by vegetation.

Until nitrous oxide is shown to be accessible to plants, any further notice of it is unnecessary in a treatise of this kind.

Hydrochloric Acid Gas, HCl, whose properties have been described in How Crops Grow, p. 118, is found in minute quantity in the air over salt marshes. It doubtless proceeds from the decomposition of the chloride of magnesium of sea-water. Sprengel has surmised its exhalation by sea-shore plants. It is found in the air near soda-works, being a product of the manufacture, and is destructive to vegetation.

Sulphurous Acid, SO_2, and **Sulphydric Acid**, H_2S, (see H. C. G., p. 115,) may exist in the atmosphere as local emanations. In large quantities, as when escaping from smelting works, roasting heaps, or manufactories, they often prove destructive to vegetation. In contact with air they quickly suffer oxidation to sulphuric acid, which, dissolving in the water of rains, etc., becomes incorporated with the soil.

Organic Matters of whatever sort that escape as vapor into the atmosphere and are there recognized by their odor, are rapidly oxidized and have no direct influence upon vegetation, so far as is now known.

Suspended Solid Matters in the Atmosphere.—The solid matters which are raised into the air by winds in the form of dust, and are often transported to great heights and distances, do not properly belong to the atmosphere, but to the soil. Their presence in the air explains the growth of certain plants (*air-plants*) when entirely disconnected from the soil, or of such as are found in pure sand or on the surface of rocks, incapable of performing the functions of the soil, except as dust accumulates upon them.

Barral announced in 1862 (*Jour. d'Ag. pratique*, p. 150) the discovery of phosphoric acid in rain-water. Robinet and Luca obtained the same result with water gathered near the surface of the earth. The latter found, however, that rain, collected at a height of 60 or more feet above the ground, was free from it.

§ 10.

RECAPITULATION OF THE ATMOSPHERIC SUPPLIES OF FOOD TO CROPS.

Oxygen, whether required in the free state to effect chemical changes in the processes of organization, or in combination (in *carbonic acid*) to become an ingredient of the plant, is superabundantly supplied by the atmosphere.

Carbon.—The *carbonic acid* of the atmosphere is a source of this element sufficient for the most rapid growth, as is abundantly demonstrated by the experiments in water culture, made by Nobbe and Siegert, and by Wolff, (H. C. G., p. 170), in which oat and buckwheat plants were brought to more than the best agricultural development, with no other than the atmospheric supply of carbon.

Hydrogen is adequately supplied to crops by *water*, which equally belongs to the Atmosphere and the Soil, although it enters the plant chiefly from the latter.

Nitrogen exists in immense quantities in the atmosphere, and we may regard the latter as the primal source of this element to the organic world. In the atmosphere, however, nitrogen exists for the most part in the free state, and is, as such, so we must believe from existing evidence, unassimilable by crops. Its assimilable compounds, *ammonia* and *nitric acid*, occur in the atmosphere, but in proportions so minute, as to have no influence on vegetable growth directly appreciable by the methods of investigation hitherto employed, unless they are collected and concentrated by rain and dew.

The subjoined Table gives a summary of the amount of nitrogen annually brought down in rain, snow, etc., upon an acre of surface, according to the determinations hitherto made in England and Prussia.

AMOUNT OF ASSIMILABLE NITROGEN ANNUALLY BROUGHT DOWN BY THE ATMOSPHERIC WATERS.

Locality.	Year.	Nitrogen per Acre.	Water per Acre.
Rothamstead, Southern England	1855*	6.63 lbs.	6,633,220 lbs.
" "	1856*	8.31 "	6,160,510 "
Kuschen, Province Posen, Prussia	1864-5†	1.86 "	2,680,086 "
" "	1865-6†	2.50 "	4,008,491 "
Insterburg, near Königsberg, "	1864-5†	5.49 "	6,222,461 "
" "	1865-6†	6.81 "	5,383,478 "
Regenwalde, near Stettin, "	1864-5†	15.09 "	5,313,562 "
" "	1865-6†	10.38 "	4,358,053 "
Ida-Marienhütte, near Breslau, Silesia, "	1865*	11.83 "	4,877,515 "
Proskau, Silesia, "	1864-5‡	20.91 "	4,031,782 "
Dahme, Province Brandenburg, "	1865*	6.66 "	3,868,646 "
Average		8.76 lbs.	4,867.075 lbs.

* From Jan. to Jan. † From Apr. to Apr. ‡ From May to May.

Direct Atmospheric Supply of Nitrogen Insufficient for Crops.—To estimate the adequacy of these atmospheric supplies of assimilable nitrogen, we may compare their amount with the quantity of nitrogen required in the composition of standard crops, and with the quantity contained in appropriate applications of nitrogenous fertilizers.

The average atmospheric supply of nutritive nitrogen in rain, etc., for 12 months, as above given, is much less than is necessary for ordinary crops. According to Dr. Anderson, the nitrogen in a crop of 28 bushels of wheat and 1 (long) ton 3 cwt. of straw, is 45¼ lbs.; that in 2½ tons of meadow hay is 56 lbs. The nitrogen in a crop of clover hay of 2½ (long) tons is no less than 108 lbs. Obviously, therefore, the atmospheric waters alone are incapable of furnishing crops with the quantity of nitrogen they require.

On the other hand, the atmospheric supply of nitrogen by rain, etc., is not inconsiderable, compared with the amount of nitrogen, which often forms an effective manuring. Peruvian guano and nitrate of soda (Chili saltpeter) each contain about 15 per cent of nitrogen. The nitrogen of rain, estimated by the average above given, viz., 8¾ lbs., corresponds to 58 lbs. of these fertilizers. 200 lbs. of guano is for most field purposes a sufficient application, and 400 lbs. is a large manuring. In Great Britain, where nitrate of soda is largely employed as a fertilizer, 112 lbs. of this substance is an ordinary dressing, which has been known to double the grass crop.

We notice, however, that the amount of nitrogen supplied in the atmospheric waters is quite variable, as well for different localities as for different years, and for different periods of the year. At Kuschen, but 2–2½ lbs. were brought down against 21 lbs. at Proskau. At Regenwalde the quantity was 15 lbs. in 1864–5, but the next year it was nearly 30 per cent less. In 1855, at Rothamstead, the greatest rain supply of nitrogen was in July, amounting to 1¼ lbs., and in October nearly as much more was brought down; the least fell in January. In 1856 the largest amount, 2¼ lbs., fell in May; the next, 1 lb., in

April; and the least in March. At Ida-Marienhütte, Kuschen, and Regenwalde, in 1865-6, nearly half the year's atmospheric nitrogen came down in summer; but at Insterburg only 30 per cent fell in summer, while 40 per cent came down in winter.

The nitrogen that is brought down in winter, or in spring and autumn, when the fields are fallow, can be counted upon as of use to summer crops only so far as it remains in the soil in an assimilable form. It is well known that, in general, much more water evaporates from cultivated fields during the summer than falls upon them in the same period; while in winter, the water that falls is in excess of that which evaporates. But how much of the winter's fall comes to supply the summer's evaporation, is an element of the calculation likely to be very variable, and not as yet determined in any instance.

We conclude, then, that the direct atmospheric supply of assimilable nitrogen, though not unimportant, is insufficient for crops.

We must, therefore, look to the soil to supply a large share of this element, as well as to be the medium through which the assimilable atmospheric nitrogen chiefly enters the plant.

The Other Ingredients of the Atmosphere, so far as we now know, are of no direct significance in the nutrition of agricultural plants. Indirectly, atmospheric ozone has an influence on the supplies of nitric acid, a point we shall recur to in a full discussion of the question of the Supplies of Nitrogen to Vegetation, in a subsequent chapter.

§ 11.

ASSIMILATION OF ATMOSPHERIC FOOD.

Boussingault has suggested the very probable view that the first process of assimilation in the chlorophyll cells of the leaf,—where, under the solar influence, carbonic acid is absorbed and decomposed, and a nearly equal volume of oxygen is set free,—consists in the simultaneous deoxidation of carbonic acid and of water, whereby the former is reduced to carbonic oxide with loss of half its oxygen, and the latter to hydrogen with loss of all its oxygen, viz.:

$$\underset{acid}{Carbonic} + Water. = \underset{oxide}{Carbonic} + \underset{gen.}{Hydro\text{-}} + \underset{gen.}{Oxy\text{-}}$$

$$CO_2 + H_2O = CO + H_2 + O_2$$

In this reaction the oxygen set free is identical in bulk with the carbonic acid involved, and the residue retained in the plant, COH_2, multiplied by 12, would give 12 molecules of carbonic oxide and 24 atoms of hydrogen, which, chemically united, might constitute either glucose or levulose, $C_{12}H_{24}O_{12}$, from which by elimination of H_2O would result cane sugar and Arabic acid, while separation of $2H_2O$ would give cellulose and the other members of its group.

Whether the real chemical process be this or a different and more complicated one is at present a matter of vague probability. It is, notwithstanding, evident that this reaction expresses one of the principal results of the assimilation of Carbon and Hydrogen in the foliage of plants.

§ 12.

The following Tabular View may usefully serve the reader as a recapitulation of the chapter now finished.

TABULAR VIEW OF THE RELATIONS OF THE ATMOSPHERIC INGREDIENTS TO THE LIFE OF PLANTS.

Absorbed by Plants.
- OXYGEN, by roots, flowers, ripening fruit, and by all growing parts.
- CARBONIC ACID, by foliage and green parts, but only in the light.
- AMMONIA, as carbonate, by foliage, probably at all times.
- WATER, as liquid, through the roots.
- NITROUS ACID } united to ammonia, and dissolved in water through the roots.
- NITRIC ACID }
- OZONE } uncertain.
- MARSH GAS }

Not absorbed by Plants.
{ NITROGEN.
{ WATER in state of vapor.

Exhaled by Plants.
{ OXYGEN, } by foliage and green parts, but only in the
{ OZONE? } light.
{ MARSH GAS in traces by aquatic plants?
{ WATER, *as vapor*, from surface of plant at all times.
{ CARBONIC ACID, from the growing parts at all times.

CHAPTER II.

THE ATMOSPHERE AS PHYSICALLY RELATED TO VEGETATION.

§ 1.

MANNER OF ABSORPTION OF GASEOUS FOOD BY THE PLANT.

Closing here our study of the atmosphere considered as a source of the food of plants, we still need to remark somewhat upon the physical properties of gases in relation to vegetable life; so far, at least, as may give some idea of the means by which they gain access into the plant.

Physical Constitution of the Atmosphere.—That the atmosphere is a *mixture* and not a chemical combination of its elements is a fact so evident as scarcely to require discussion. As we have seen, the proportions which subsist among its ingredients are not uniform, although they are ordinarily maintained within very narrow limits of variation. This is a sufficient proof that it is a mixture. The remarkable fact that very nearly the same relative quantities of Oxygen, Nitrogen, and Carbonic Acid, steadily exist in the atmosphere is due to the even balance which obtains between growth and decay, between life and death. The equally remarkable fact that the gases which compose the atmosphere are uniformly mixed together without regard to their specific gravity, is but one result of a law of nature which we shall immediately notice.

Diffusion of Gases.—Whenever two or more gases are brought into contact in a confined space, they instantly begin to intermingle, and continue so to do until, in a longer or shorter time, they are both equally diffused throughout the room they occupy. If two bottles, one filled with carbonic acid, the other with hydrogen, be connected by a tube no wider than a straw, and be placed so that the heavy carbonic acid is below the fifteen times lighter hydrogen, we shall find, after the lapse of a few hours, that the two gases have mingled somewhat, and in a few days they will be in a state of uniform mixture. On closer study of this phenomenon it has been discovered that gases diffuse with a rapidity proportioned to their lightness, the relative diffusibility being nearly in the inverse ratio of the square roots of their specific gravities. By interposing a porous diaphragm between two gases of different densities, we may visibly exhibit the fact of their ready and unequal diffusion. For this purpose the diaphragm must offer a partial resistance to the movement of the gases. Since the lighter gas passes more rapidly into the denser than the reverse, the space on one side of the membrane will be overfilled, while that on the other side will be partially emptied of gas.

In the accompanying figure is represented a long glass tube, b, widened above into a funnel, and having cemented upon this an inverted cylindrical cup of unglazed porcelain, a. The funnel rests in a round aperture made in the horizontal arm of the support, while the tube below dips beneath the surface of some water contained in the wineglass. The porous cup, funnel, and tube, being occupied with common air, a glass bell, c, is filled with hydrogen gas and placed over the cap, as shown in the figure. Instantly, bubbles begin to escape rapidly from the bottom of the tube through the water of the wine-glass, thus demonstrating that hydrogen passes into the cup faster than air can escape outwards through its pores. If the bell be removed, the cup is at once bathed again externally in common air, the light hydrogen floating instantly upwards, and now the water begins to rise in the tube in consequence of the return to the outer atmosphere of the hydrogen which before had diffused into the cup.

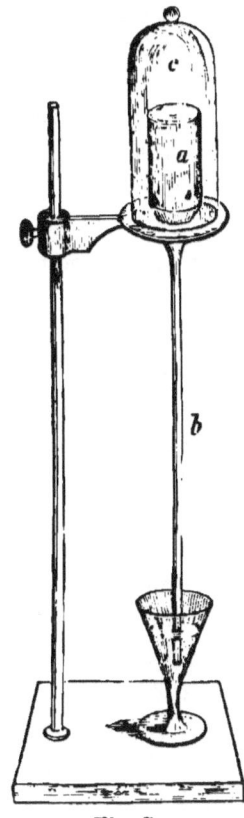

Fig. 7.

It is the perpetual action of this diffusive tendency which maintains the atmosphere in a state of such uniform mixture that accurate analyses of it give for oxygen and nitrogen almost identical figures, at all times of the day, at all seasons, all altitudes, and all situations, except near the central surface of large bodies of still water. Here, the fact that oxygen is more largely absorbed by water than nitrogen, diminishes by a minute amount the usual proportion of the former gas.

If in a limited volume of a mixture of several gases a solid or liquid body be placed, which is capable of chemically uniting with, or otherwise destroying the aeriform condition of one of the gases, it will at once absorb those particles of this gas which lie in its immediate vicinity, and thus disturb the uniformity of the remaining mixture. Uniformity at once tends to be restored by diffusion of a portion of the unabsorbed gas into the space that has been deprived of it, and thus the absorption and the diffusion keep pace with each other until all the absorbable air is removed from the gaseous mixture, and condensed or fixed in the absorbent.

In this manner, a portion of the atmosphere enclosed in a large glass vessel may be perfectly freed from watery vapor and carbonic acid by a small fragment of caustic potash. By standing over sulphuric acid, ammonia is taken from it; a piece of phosphorus will in a few hours absorb all its oxygen, and an ignited mass of the rare metal titanium will remove its nitrogen.

Osmose of Gases.—By this expression is understood the passage of gaseous bodies through *membranes whose pores are too small to be discoverable by optical means*, such as the imperforate wall of the vegetable cell, the green cuticle of the plant where not interrupted by *stomata*, vegetable parchment, India rubber, and animal membranes, like bladder and similar visceral integuments.

If a bottle filled with air have a thin sheet of India rubber, or a piece of moist bladder tied over its mouth and then be placed within a bell of hydrogen, evidence is at once had that gases penetrate the membrane, for it swells outwards, and may even burst by the pressure of the hydrogen that rapidly accumulates in the bottle.

Gaseous Osmose is Diffusion Modified by the Influence of the Membrane.—The rapidity of osmose* is of course influenced by the thickness of the membrane, and the character of its pores. An adhesion between the membrane and the gases would necessarily increase their rate of penetration. In case the membrane should attract or have adhesion for one gas and not for another, complete separation of the two might be accomplished, and in proportion to the difference existing between two gases as regards adhesion for a given membrane, would be the degree to which such gases would be separated from each

---

* The osmose of liquids is discussed in detail in "How Crops Grow," p. 354.

other in penetrating it. In case a membrane is moistened with water or other liquid, or by a solution of solid matters, this would still further modify the result.

Absorption of Gases by the Plant.—A few words will now suffice to apply these facts to the absorption of the nutritive gases by vegetation. The foliage of plants is freely permeable to gases, as has been set forth in "How Crops Grow," p. 289. The cells, or some portions of their contents, absorb or condense carbonic acid and ammonia in a similar way, or at least with the same effect, as potash absorbs carbonic acid. As rapidly as these bodies are removed from the atmosphere surrounding or occupying the cells, they are re-supplied by diffusion from without; so that although the quantities of gaseous plant-food contained in the air are, relatively considered, very small, they are by this grand natural law made to flow in continuous streams toward every growing vegetable cell.

DIVISION II.

THE SOIL AS RELATED TO VEGETABLE PRODUCTION.

CHAPTER I.

INTRODUCTORY.

For the Husbandman the Soil has this paramount importance, that it is the home of the roots of his crops and the exclusive theater of his labors in promoting their growth. Through it alone can he influence the amount of vegetable production, for the atmosphere, and the light and heat of the sun, are altogether beyond his control. Agriculture is the culture of the field. The value of the field lies in the quality of its soil. No study can have a grander material significance than the one which gives us a knowledge of the causes of fertility and barrenness, a knowledge of the means of economizing the one and overcoming the other, a knowledge of those natural laws which enable the farmer so to modify and manage his soil that all the deficiencies of the atmosphere or the vicissitudes of climate cannot deprive him of a suitable reward for his exertions.

The atmosphere and all extra-terrestrial influences that affect the growth of plants are indeed in themselves beyond our control. We cannot modify them in kind or amount; but we can influence their subserviency to our purposes through the medium of the soil by a proper understanding of the characters of the latter.

The General Functions of the Soil are of three kinds:

1. The ashes of the plant whose nature and variations have been the subject of study in a former volume (II. C. G., pp. 111–201,) are exclusively derived from the soil. The latter is then concerned in the most direct manner with the nutrition of the plant. The substances which the plant acquires from the soil, so far as they are nutritive, may be collectively termed *soil-food*.

2. The soil is a mechanical support to vegetation. The roots of the plant penetrate the pores of the soil in all directions sidewise and downward from the point of their junction with the stem, and thus the latter is firmly braced to its upright position if that be natural to it, and in all cases is fixed to the source of its supplies of ash-ingredients.

3. By virtue of certain special (physical) qualities to be hereafter enumerated, the soil otherwise contributes to the well-being of the plant, tempering and storing the heat of the sun which is essential to the vital processes; regulating the supplies of food, which, coming from itself or from external sources, form at any one time but a minute fraction of its mass, and in various modes ensuring the co-operation of the conditions which must unite to produce the perfect plant.

Variety of Soils.—In nature we observe a vast variety of soils, which differ as much in their agricultural value as they do in their external appearance. We find large tracts of country covered with barren, drifting sands, on whose arid bosom only a few stunted pines or shriveled grasses find nourishment. Again there occur in the highlands of Scotland and Bavaria, as well as in Prussia, and other temperate countries, enormous stretches of moorland, bearing a nearly useless growth of heath or moss. In Southern Russia occurs a vast tract, two hundred millions of acres in extent, of the *tschornosem*, or black earth, which is remarkable for its extraordinary and persistent fertility. The prairies of our own West, the bottom lands of the Scioto and other rivers of Ohio, are other examples of peculiar soils; while on every farm, almost, may be found numerous gradations from clay to sand, from vegetable mould to gravel—gradations in color, consistence, composition, and productiveness.

CHAPTER II.

ORIGIN AND FORMATION OF SOILS.

Some consideration of the origin of soils is adapted to assist in understanding the reasons of their fertility. Geological studies give us reasons to believe that what is now soil was once, in chief part, solid rock. We find in nearly all soils fragments of rock, recognizable as such by the eye, and by help of the microscope it is often easy to perceive that those portions of the soil which are impalpable to the feel are only minuter grains of the same rock.

Rocks are aggregates or mixtures of certain *minerals*.

Minerals, again, are chemical compounds of various elements.

We have therefore to consider:

I. The Chemical Elements of Rocks.

II. The Mineralogical Elements of Rocks.

III. The Rocks themselves—their Kinds and Special Characters.

IV. The Conversion of Rocks into Soils; to which we may add:

V. The Incorporation of Organic Matter with Soils.

§ 1.

THE CHEMICAL ELEMENTS OF ROCKS.

The chemical elements of rocks, i. e., the constituents of the minerals which go to form rocks, include all the simple bodies known to science. Those, which, from their universal distribution and uses in agriculture, concern us immediately, are with one exception the same that have been noticed in a former volume as composing the ash of agricultural plants, viz., Chlorine, Sulphur, Carbon, Silicon, Potassium, Sodium, Calcium, Magnesium, Iron, and Manganese. The description given of these elements and of their most important compounds in "How Crops Grow" will suffice. It is only needful to notice further a single element.

Aluminum, *Symbol* Al., *At. wt.* 27.4, is a bluish silver-white metal, characterized by its remarkable lightness, having about the specific gravity of glass. It is now manufactured on a somewhat large scale in Paris and Newcastle, and is employed in jewelry and ornamental work. It is prepared by a costly and complex process invented by Prof. Deville, of Paris, in 1854, which consists essentially in decomposing chloride of aluminum by metallic sodium, at a high heat, chloride of sodium (common salt) and metallic aluminum being produced, as shown by the equation, $Al_2 Cl_6 + 6 Na = 6 NaCl + 2 Al$.

By combining with oxygen, this metal yields but one oxide, which, like the highest oxide of iron, is a sesquioxide, viz.:

Alumina, $Al_2 O_3$, Eq. 102.8.—When alum (double sulphate of alumina and potash) is dissolved in water and ammonia added to the solution, a white gelatinous body separates, which is alumina combined with water, $Al_2 O_3$, $3 H_2 O$. By drying and strongly heating this hydrated alumina, a white powder remains, which is pure alumina.

In nature alumina is found in the form of *emery*. The sapphire and ruby are finely colored crystallized varieties of alumina, highly prized as gems.

Hydrated alumina dissolves in acids, yielding a numerous class of salts, of which the sulphate and acetate are largely employed in dyeing and calico-printing. The sulphate of alumina and potash is familiarly known under the name of alum, with which all are acquainted. Other compounds of alumina will be noticed presently.

§ 2.

MINERALOGICAL ELEMENTS OF ROCKS.

The mineralogical elements or minerals* which compose rocks are very numerous.

But little conception can be gained of the appearance of a mineral from a description alone. Actual inspection of the different varieties is necessary to enable one to recognize them. The teacher should be provided with a collection to illustrate this subject. The true idea of their composition and use in forming rocks and soils may be gathered quite well, however, from the written page. For minute information concerning them, see Dana's Manual of Mineralogy. We shall notice the most important.

Quartz.—Chemically speaking, this mineral is anhydrous silica—silicic acid—a compound of silicon and oxygen, Si O_2. It is one of the most abundant substances met with on the earth's surface. It is found in nature in six-sided crystals, and in irregular masses. It is usually colorless, or white, irregular in fracture, glassy in luster. It is very hard, readily scratching glass. (See H. C. G., p. 120.)

Feldspar (field-spar) is, next to quartz, the most abund-

---

* The word mineral, or mineral "species," here implies a definite chemical compound of natural occurrence.

ant mineral. *It is a compound of silica with alumina, and with one or more of the alkalies, and sometimes with lime.* Mineralogists distinguish several species of feldspar according to their composition and crystallization. Feldspar is found in crystals or crystalline masses usually of a white, yellow, or flesh color, with a somewhat pearly luster on the smooth and level surfaces which it presents on fracture. It is scratched by, and does not scratch quartz.

In the subjoined Table are given the mineralogical names and analyses of the principal varieties of feldspar. Accompanying each analysis is its locality and the name of the analyst.

	ORTHOCLASE. Common or potash feldspar. New Rochelle, N. Y. S. W. Johnson.	ALBITE. Soda feldspar. Unionville, Pa. M. C. Weld.	OLIGOCLASE. Soda-lime feldspar. Haddam, Conn. G. J. Brush.	LABRADORITE. Lime-soda feldspar. Drummond, C. W. T. S. Hunt.
Silica,	64.23	66.86	64.26	54.70
Alumina,	20.42	21.89	21.90	29.80
Potash,	12.47	—	0.50	0.33
Soda,	2.62	8.78	9.99	2.44
Lime,	trace	1.79	2.15	11.42
Magnesia,	—	0.48	—	—
Oxide of iron,	trace	—	—	0.36
Water,	0.21	0.18	0.29	0.40

Mica is, perhaps, next to feldspar, the most abundant mineral. There are three principal varieties, viz.: Muscovite, Phlogopite, and Biotite. They are silicates of alumina with potash, magnesia, lime, iron, and manganese.

Mica bears the common name "isinglass." It readily splits into thin, elastic plates or leaves, has a brilliant luster, and a great variety of colors,—white, yellow, brown, green, and black. *Muscovite*, or muscovy glass, is sometimes found in transparent sheets of great size, and is used in stove-doors and lamp-chimneys. It contains much alumina, and potash, or soda, and the black varieties oxide of iron.

Phlogopite and *Biotite* contain a large percentage of magnesia, and often of oxide of iron.

The following analyses represent these varieties.

	MUSCOVITE.		PHLOGOPITE.		BIOTITE.	
	Litchfield, Conn.	Mt. Leinster, Ireland.	Edwards, N. Y.	N. Burgess, Canada.	Putnam Co., N. Y.	Siberia.
	Smith & Brush.	Haughton.	W.J.Craw.	T.S.Hunt.	Smith & Brush.	H. Rose.
Silica,	44.60	44.64	40.36	40.97	39.62	40.00
Alumina,	36.23	30.18	16.45	18.56	17.35	12.67
Oxide of iron,	1.31	6.35	trace	—	5.40	19.03
Oxide of manganese,	—	—	—	—	—	0.63
Magnesia,	0.37	0.72	29.55	25.80	23.85	15.70
Lime,	0.50	—	—	—	—	—
Potash,	6.20	12.40	7.23	8.26	8.95	5.61
Soda,	4.10	—	4.94	1.08	1.01	—
Water,	5.26	5.32	0.95	1.00	1.41	—

Variable Composition of Minerals.—We notice in the micas that two analyses of the same species differ very considerably in the proportion, and to some extent in the kind, of their ingredients. Of the two muscovites the first contains 6°|₀ more of alumina than the second, while the second contains 5°|₀ more of oxide of iron than the first. Again, the second contains 12.4°|₀ of potash, but no soda and no lime, while the first reveals on analysis 4°|₀ of soda and 0.5°|₀ of lime, and contains correspondingly less potash. Similar differences are remarked in the other analyses, especially in those of Biotite.

In fact, of the analyses of more than 50 micas which are given in mineralogical treatises, scarcely any two perfectly agree. The same is true of many other minerals, especially of the amphiboles and pyroxenes presently to be noticed. In accordance with this variation in composition we notice extraordinary diversities in the color and appearance of different specimens of the same mineral.

This fact may appear to stand in contradiction to the statement above made that these minerals are *definite combinations*. In the infancy of mineralogy great perplexity arose from the numerous varieties of minerals that were found—varieties that agreed together in certain characteristics, but widely differed in others.

Isomorphism.—In 1830, Mitscherlich, a Prussian philosopher, discovered that a number of the elementary bodies are capable of *replacing each other in combination*, from the fact of their natural crystalline form being identical; they being, as he termed it, *isomorphous*, or *of like shape*. Thus, magnesia, lime, protoxide of iron, protoxide of manganese, which are all *protoxide-bases*, form one group, each of whose members may take the place of the other. Alumina (Al_2O_3) and oxide of iron (Fe_2O_3) belong to another group of *sesquioxide-bases*, one of which may replace the other; while in certain combinations silica and alumina replace each other as *acids*.

These replacements, which may take place indefinitely within certain limits, thus may greatly affect the composition without altering the constitution of a mineral. Of the mineral *amphibole*, for example, there are known a great number of varieties; some pure white in color, containing, in addition to silica, magnesia and lime; others pale green, a small portion of magnesia being replaced by protoxide of iron; others black, containing alumina in place of a portion of silica, and with oxides of iron and manganese in large proportion. All these varieties of amphibole, however, admit of one expression of their constitution, for the amount of *oxygen* in the bases, no matter what they are, or what their proportions, bears a constant relation to the *oxygen* of the silica (and alumina) they contain, the ratio being 1 : 2.

If the protoxides be grouped together under the general symbol MO (metallic protoxide,) the composition of the amphiboles may be expressed by the formula $MO\ SiO_2$.

In *pyroxene* the same replacements of protoxide-bases on the one hand, and of silica and alumina on the other, occur in extreme range. (See analyses, p. 112.) The general formula which includes all the varieties of pyroxene is the same as that of amphibole. The distinction of amphibole from pyroxene is one of *crystallization*.

We might give in the same style formulæ for all the minerals noticed in these pages, but for our purposes this is unnecessary.

Amphibole is an abundant mineral often met with in distinct crystals or crystalline and fibrous masses, varying in color from pure white or gray (*tremolite, asbestus*), light green (*actinolite*), grayish or brownish green (*anthophyllite*), to dark green and black (*hornblende*), according as it contains more or less oxides of iron and manganese. It is a silicate of magnesia and lime, or of magnesia and protoxide of iron, with more or less alkalies.

	White. Gouverneur, N. Y. Rammelsberg.	Gray. Lanark, Canada. T. S. Hunt.	Ash-gray. Cummington, Mass. Smith & Brush.	Black. Brevig, Norway. Plantamour.	Leek-green. Waldheim, Saxony. Kuop.
Silica,	57.40	55.30	50.74	46.37	58.71
Magnesia,	24.69	22.50	10.31	5.88	10.01
Lime,	13.89	13.36	trace	5.91	11.53
Protoxide of iron,	1.36	6.30	33.14	24.38	5.65
Protoxide of manganese,	—	trace	1.77	2.07	—
Alumina,	1.38	0.40	0.89	3.41	1.52
Soda,	—	0.80	0.54	7.79	12.38
Potash,	—	0.25	trace	2.96	—
Water,	0.40	0.30	3.01	—	0.50

Pyroxene is of very common occurrence, and considerably resembles hornblende in colors and in composition.

	White. Ottawa, Canada. T. S. Hunt.	Gray-White. Bathurst, Canada. T. S. Hunt.	Green. Lake Champlain. Seybert.	Black. Orange Co., N. Y. Smith & Brush.	Black. Wetterau. Gmelin.
Silica,	54.50	51.50	50.38	39.30	56.80
Magnesia,	18.14	17.69	6.83	2.98	5.05
Lime,	25.87	23.80	19.33	10.39	4.85
Protoxide of iron,	1.98	—	20.40	30.40	12.06
Sesquioxide of iron,	—	0.35	—	—	—
Protoxide of manganese,	—	—	trace	0.67	3.72
Alumina,	—	6.15	1.83	9.78	15.32
Soda,	—	—	—	1.66	3.14
Potash,	—	—	—	2.48	0.34
Water,	0.40	1.10	—	1.95	—

ORIGIN AND FORMATION OF SOILS. 113

Chlorite is a common mineral occurring in small scales or plates which are *brittle*. It is soft, usually exists in masses, rarely crystallized, and is very variable in color and composition, though in general it has a grayish or brownish-green color, and contains magnesia, alumina, and iron, united with silica. See analysis below.

Leucite is an anhydrous silicate of alumina found chiefly in volcanic rocks. It exists in white, hard, 24-sided crystals. It is interesting as being formed at a high heat in melted lava, and as being the first mineral in which potash was discovered (by Klaproth, in 1797). See analysis below.

Kaolinite is a hydrous silicate of alumina, which is produced by the slow decomposition of feldspar under the action of air and water at the usual temperature. Formed in this way, in a more or less impure state, it constitutes the mass of white porcelain clay or kaolin, which is largely used in making the finer kinds of pottery. It appears in white or yellowish crystalline scales of a pearly luster, or as an amorphous translucent powder of extreme fineness. Ordinary clay is a still more impure kaolinite.

	CHLORITE.	LEUCITE.	KAOLINITE.	
	Steele Mine, N. C. Genth.	Vesuvius, Eruption of 1857. Rammelsberg.	Summit Hill, Pa. S. W. Johnson.	Chaudiere Falls, Canada. T. S. Hunt.
Silica,	21.90	57.24	45.93	46.05
Alumina,	21.77	22.96	39.81	38.37
Sesquioxide of iron,	4.60	—	—	—
Protoxide of iron,	24.21	—	—	—
Protoxide of manganese,	1.15	—	—	—
Magnesia,	12.78	—	—	0.63
Lime,	—	0.91	—	0.61
Soda,	—	0.93	—	—
Potash,	—	18.61	—	—
Water,	10.59	—	14.02	14.00

Talc is often found in pale-green, flexible, inelastic scales or leaves, but much more commonly in compact gray masses, and is then known as soapstone. It is very soft, has a greasy feel, and in composition is a hydrous silicate of magnesia. See analysis.

Serpentine is a tough but soft massive mineral, in color usually of some shade of green. It forms immense beds in New England, New York, Pennsylvania, etc. It is also a hydrous silicate of magnesia. See analysis.

Chrysolite is a silicate of magnesia and iron, which is found abundantly in lavas and basaltic rocks. It is a hard, glassy mineral, usually of an olive or brown-green color. See analysis below.

	TALC. Bristol, Conn. Dr. Lummis.	SERPENTINE. New Haven, Conn. G. J. Brush.	CHRYSOLITE. Bolton, Mass. G. J. Brush.
Silica,	64.00	44.05	40.94
Alumina,	—	—	0.27
Protoxide of iron,	4.75	2.53	4.37
Magnesia,	27.47	39.24	50.84
Lime,	—	—	1.20
Water,	4.30	13.49	3.28

Zeolites.—Under this general name mineralogists are in the habit of including a number of minerals which have recently acquired considerable agricultural interest, since they represent certain compounds which we have strong reasons to believe are formed in and greatly influence the properties of soils. They are hydrous silicates of alumina or lime, and alkali, and are remarkable for the ease with which they undergo decomposition under the influence of weak acids. We give here the names and composition of the most common zeolites. Their special significance will come under notice hereafter. We may add that while they all occur in white or red crystallizations, often of great beauty, they likewise exist in a state of division so minute that the eye cannot recognize them, and thus form a large share of certain rocks, which, by their disintegration, give origin to very fertile soils

ORIGIN AND FORMATION OF SOILS.

	ANALCIME. Lake Superior. C. T. Jackson.	CHABASITE. Nova Scotia. Rammelsberg.	NATROLITE. Bergen Hill, N. J. Brush.	SCOLECITE. Ghant's Tunnel, India. P. Collier.	THOMSONITE. Magnet Cove, Ark. Smith & Brush
Silica,	53.40	52.14	47.31	45.80	36.85
Alumina,	22.40	19.14	26.77	25.55	29.42
Potash,	—	0.98	0.35	0.30	—
Soda,	8.52	0.71	15.44	0.17	3.91
Lime,	3.00	7.84	0.41	13.97	13.95
Magnesia,	—	—	—	—	—
Sesquioxide of iron,	—	—	—	—	1.55
Water,	9.70	19.19	9.84	14.28	13.80

	STILBITE. Nova Scotia. S. W. Johnson.	APOPHYLLITE. Lake Superior. J. L. Smith.	PECTOLITE. Bergen Hill. J. D. Whitney.	LAUMONTITE. Phippsburgh, Me. Dufrénoy.	LEONHARDITE. Lake Sup'r Barnes.
Silica,	57.63	52.08	55.66	51.98	55.01
Alumina,	16.17	—	1.45	21.12	22.34
Potash,	—	4.93	—	—	—
Soda,	1.55	—	8.89	—	—
Lime,	8.08	25.30	32.86	11.71	10.64
Water,	16.07	15.92	2.96	15.05	11.93

Calcite, or *Carbonate of Lime,* $CaO\ CO_2$, exists in nature in immense quantities as a mineral and rock. Marble, chalk, coral, limestone in numberless varieties, consist of this substance in a greater or less state of purity.

Magnesite, or *Carbonate of Magnesia,* $MgO\ CO_2$, occurs to a limited extent as a white massive or crystallized mineral, resembling carbonate of lime.

Dolomite, $CaO\ CO_2 + MgO\ CO_2$, is a compound of carbonate of lime with carbonate of magnesia in variable proportions. It is found as a crystallized mineral, and is a very common rock, many so-called marbles and limestones consisting of or containing this mineral.

Gypsum, or *Hydrous Sulphate of Lime,* $CaO\ SO_3 + H_2O$, is a mineral that is widely distributed and quite abundant in nature. When "boiled" to expel the water it is Plaster of Paris.

Pyrites, or *Bisulphide of Iron,* $Fe\ S_2$, a yellow shining mineral often found in cubic or octahedral crystals, and frequently mistaken for gold (hence called fool's gold), is of almost universal occurrence in small quantities. Some forms of it easily oxidize when exposed to air, and furnish the green-vitriol (sulphate of protoxide of iron) of commerce.

Apatite and **Phosphorite.**—These names are applied to the native phosphate of lime, which is usually combined with some chlorine and fluorine, and may besides contain other ingredients. Apatite exists in considerable quantity at Hammond and Gouverneur, in St. Lawrence Co., N. Y., in beautiful, transparent, green crystals; at South Burgess, Canada, in green crystals and crystalline masses; at Hurdstown, N. J., in yellow crystalline masses; at Krageröe, Norway, in opaque flesh-colored crystals. In minute quantity apatite is of nearly universal distribution. The following analyses exhibit the composition of the principal varieties.

	Krageröe, Norway. Voelcker.	Hurdstown, New Jersey. J. D. Whitney.
Lime,	53.84	53.37
Phosphoric acid,	41.25	42.23
Chlorine,	4.10	1.02
Fluorine,*	1.23 ?	?
Oxide of iron,	0.29	trace
Alumina,	0.38	
Potash and soda,	0.17	
Water,	0.42	

Phosphorite is the usual designation of the non-crystalline varieties.

Apatite may be regarded as a mixture in indefinite proportions of two isomorphous compounds, *chlorapatite* and *fluorapatite*, neither of which has yet been found pure in nature, though they have been produced artificially.

---

* Fluorine was not determined in these analyses. The figures given for this element are calculated (by Rammelsberg), and are probably not far from the truth.

These substances are again compounds of phosphate of lime, $3\,CaO\,P_2O_5$, with chloride of calcium, $Ca\,Cl_2$, or fluoride of calcium, $Ca\,Fl_2$, respectively.

§ 3.

ROCKS—THEIR KINDS AND CHARACTERS.

The Rocks which form the solid (unbroken) mass of the earth are sometimes formed from a single mineral, but usually contain several minerals in a state of more or less intimate mixture.

We shall briefly notice those rocks which have the greatest agricultural importance, on account of their common and wide-spread occurrence, and shall regard them principally from the point of view of their *chemical composition*, since this is chiefly the clue to their agricultural significance. Some consideration of the *origin* of rocks, as well as of their *structure*, will also be of service.

Igneous Rocks.—A share of the rocks accessible to our observation are plainly of *igneous origin*, i. e., their existing form is the one they assumed on cooling down from a state of fusion by heat. Such are the lavas that flow from volcanic craters.

Sedimentary Rocks.—Another share of the rocks are of *aqueous origin*, i. e., their materials have been deposited from water in the form of mud, sand, or gravel, the loose sediment having been afterwards cemented and consolidated to rock. The rocks of aqueous origin are also termed *sedimentary rocks*.

Metamorphic Rocks.—Still another share of the rocks have resulted from the alteration of aqueous sediments or sedimentary rocks by the effect of heat. Without suffering fusion, the original materials have been more or less converted into new combinations or new forms. Thus limestone has been converted into statuary marble, and clay into granite. These rocks, which are the result of the united action of heat and water, are termed *metamorphic* (i. e., metamorphosed) rocks.

One of the most obvious division of rocks is into *Crystalline* and *Fragmental*.

Crystalline Rocks are those whose constituents crystallized at the time the rock was formed. Here belong both the igneous and metamorphic rocks. These are often plainly crystalline to the eye, i. e., are composed of readily perceptible crystals or crystalline grains, like statuary marble or granite; but they are also frequently made up of crystals so minute, that the latter are only to be recognized by tracing them into their coarser varieties (basalt and trap.)

Fragmental Rocks are the sedimentary rocks, formed by the cementing of the *fragments* of other older rocks existing as mud, sand, etc.

THE CRYSTALLINE ROCKS may be divided into two great classes, viz., the *silicious* and *calcareous;* the first class containing silica, the latter, lime, as the predominating ingredient.

The silicious rocks fall into three parallel series, which have close relations to each other. 1. The *Granitic* series; 2. The *Syenitic* series; 3. The *Talcose* or *Magnesian* series. In all the silicious rocks *quartz* or *feldspar* is a prominent ingredient, and in most cases these two minerals are associated together. To the above are added, in the granitic series, *mica;* in the syenitic series, *amphibole* or *pyroxene;* and in the talcose series, *talc*, *chlorite*, or *serpentine*. The proportions of these minerals vary indefinitely.

THE GRANITIC SERIES
consisting principally of Quartz, Feldspar, and Mica.

Granite.— A hard, massive* rock, either finely or

---

* Rocks are *massive* when they have no tendency to split into slabs or plates

coarsely crystalline, of various shades of color, depending on the color and proportion of the constituent minerals, usually gray, grayish white, or flesh-red. In *common granite* the feldspar is *orthoclase* (potash-feldspar). A variety contains *albite* (soda-feldspar). Other kinds (less common) contain *oligoclase* and *labradorite*.

Gneiss differs from granite in containing more mica, and in having a banded appearance and schistose* structure, due to the distribution of the mica in more or less parallel layers. It is cleavable along the planes of mica into coarse slabs.

Mica-slate or *Mica-schist* contains a still larger proportion of mica than gneiss; it is perfectly schistose in structure, splitting easily into thin slabs, has a glistening appearance, and, in general, a grayish color. The coarse whetstones used for sharpening scythes, which are quarried in Connecticut and Rhode Island, consist of this mineral.

Argillite, Clay-slate, is a rock of fine texture, often not visibly crystalline, of dull or but slightly glistening surface, and having a great variety of colors, in general black, but not rarely red, green, or light gray. Argillite has usually a *slaty cleavage*, i. e., it splits into thin and smooth plates. It is extensively quarried in various localities for roofing, and writing-slates. Some of the finest varieties are used for whetstones or hones.

Other Granitic Rocks.—Sometimes *mica* is absent; in other cases the rock consists nearly or entirely of *feldspar* alone, or of *quartz* alone, or of *mica* and *quartz*. The rocks of this series often insensibly gradate into each other, and by admixture of other minerals run into numberless varieties.

---

* *Schists* or schistose rocks are those which have a tendency to break into slabs or plates from the arrangement of some of the mineral ingredients in layers.

THE SYENITIC SERIES
consisting chiefly of Quartz, Feldspar, and Amphibole.

Syenite is granite, save that amphibole takes the place of mica. In appearance it is like granite; its color is usually dark gray. Syenite is a very tough and durable rock, often most valuable for building purposes. The famous Quincy granite of Massachusetts is a syenite. *Syenitic Gneiss* and *Hornblende Schist* correspond to common Gneiss and Mica Schist, hornblende taking the place of mica.

THE VOLCANIC SERIES
consisting of Feldspar, Amphibole or Pyroxene, and Zeolites.

Diorite is a compact, tough, and heavy rock, commonly greenish-black, brownish-black, or grayish-black in color. It contains amphibole, but no pyroxene, and is an ancient lava.

Dolerite or **Trap** in the fine-grained varieties is scarcely to be distinguished from Diorite by the appearance, and is well exhibited in the Palisades of the Hudson and the East and West Rocks of New Haven. It contains pyroxene in place of amphibole.

Basalt is like dolerite, but contains grains of chrysolite. The recent lavas of volcanic regions are commonly basaltic in composition, though very light and porous in texture.

Porphyry.—Associated with basalt occur some feldspathic lavas, of which porphyry is common. It consists of a compact base of feldspar, with disseminated crystals of feldspar usually lighter in color than the mass of the rock.

Pumice is a vesicular rock, having nearly the composition of feldspar.

THE MAGNESIAN SERIES
consisting of Quartz, Feldspar and Talc, or Chlorite.

Talcose Granite differs from common granite in the substitution of talc for mica. Is a fragile and more easily decomposable rock than granite. It passes through *talcose gneiss* into

Talcose Schist, which resembles mica-schist in colors and in facility of splitting into slabs, but has a less glistening luster and a soapy feel.

Chloritic Schist resembles talcose schist, but has a less unctuous feel, and is generally of a dark green color.

Related to the above are **Steatite,** or *soapstone,*—nearly pure, granular talc; and **Serpentine rock,** consisting chiefly of *serpentine.*

The above are the more common and wide-spread silicious rocks. By the blending together of the different members of each series, and the related members of the different series, and by the introduction of other minerals into their composition, an almost endless variety of silicious rocks has been produced. Turning now to the

CRYSTALLINE CALCAREOUS ROCKS, we have

Granular Limestone, consisting of a nearly pure carbonate of lime, in more or less coarse grains or crystals, commonly white or gray in color, and having a glistening luster on a freshly broken surface. The finer kinds are employed as *monumental marble.*

Dolomite has all the appearance of granular limestone, but contains a large (variable) amount of carbonate of magnesia.

THE FRAGMENTAL OR SEDIMENTARY ROCKS are as follows:

Conglomerates have resulted from the consolidation of rather coarse fragments of any kind of rock. According to the nature of the materials composing them, they may be *granitic, syenitic, calcareous, basaltic,* etc., etc. They pass into

Sandstones, which consist of small fragments (sand), are generally *silicious* in character, and often are nearly pure quartz. The freestone of the Connecticut Valley is a *granitic* sandstone, containing fragments of feldspar and spangles of mica.

Other varieties are *calcareous, argillaceous* (*clayey*), *basaltic*, etc., etc.

Shales are soft, slaty rocks of various colors, gray, green, red, blue, and black. They consist of compacted clay. When crystallized by metamorphic action, they constitute *argillite*.

Limestones of the sedimentary kind are soft, compact, nearly lusterless rocks of various colors, usually gray, blue, or black. They are sometimes nearly pure carbonate of lime, but usually contain other substances, and are often highly impure. When containing much carbonate of magnesia they are termed *magnesian limestones*. They pass into sandstones through intermediate *calciferous sand rocks*, and into shales through *argillaceous limestones*. These impure limestones furnish the *hydraulic cements* of commerce.

§ 4.

CONVERSION OF ROCKS INTO SOILS.

Soils are broken and decomposed rocks. We find in nearly all soils fragments of rock, recognizable as such by the eye, and by help of the microscope it is often easy to perceive that those portions of the soil which are impalpable to the feel chiefly consist of minuter grains of the same rock.

Geology makes probable that the globe was once in a melted condition, and came to its present state through a process of *cooling*. By loss of heat its exterior surface solidified to a crust of solid rock, totally incapable of supporting the life of agricultural plants, being impenetrable to their roots, and destitute of all the other external characteristics of a soil.

The first step towards the formation of a soil must have been the pulverization of the rock. This has been accomplished by a variety of agencies acting through long periods of time. The causes which could produce such results are indeed stupendous when contrasted with the narrow experience of a single human life, but are really trifling compared with the magnitude of the earth itself, for the soil forms upon the surface of our globe, whose diameter is nearly 8,000 miles, a thin coating of dust, measured in its greatest accumulations not by miles, nor scarcely by rods, but by feet.

The conversion of rocks to soils has been performed, 1st, by *Changes of Temperature;* 2d, by *Moving Water or Ice;* 3d, by the *Chemical Action of Water and Air;* 4th, by the *Influence of Vegetable and Animal Life.*

1.—CHANGES OF TEMPERATURE.

The continued cooling of the globe after it had become enveloped in a solid rock-crust must have been accompanied by a contraction of its volume. One effect of this shrinkage would have been a subsidence of portions of the crust, and a wrinkling of other portions, thus producing on the one hand sea-basins and valleys, and on the other mountain ranges. Another effect would have been the cracking of the crust itself as the result of its own contraction.

The pressure caused by contraction or by mere weight of superincumbent matter doubtless led to the production of the laminated structure of slaty rocks, which may be readily imitated in wax and clay by aid of an hydraulic press. Basaltic and trap rocks in cooling from fusion often acquire a tendency to separate into vertical columns, somewhat as moist starch splits into five or six-sided fragments, when dried. These columns are again transversely jointed. The Giant's Causeway of Ireland is an illustration. These fractures and joints are, perhaps, the first occasion of the breaking down of the rocks. The fact that many rocks consist of crystalline grains of distinct minerals more or less intimately blended, is a point of weakness in their structure. The grains of quartz, feldspar, and mica, of a granite, when exposed to changes of temperature, must tend to separate from each other; because the extent to which they expand and contract by alternations of heat and cold are not absolutely equal, and because, as Senarmont has proved, the same crystal expands or contracts unequally in its different diameters.

Action of Freezing Water.—It is, however, when water insinuates itself into the slight or even imperceptible rifts thus opened, and then freezes, that the process of disintegration becomes more rapid and more vigorous. Water in the act of conversion into ice expands $\frac{1}{3}$ of its bulk, and the force thus exerted is sufficient to burst vessels of the strongest materials. In cold latitudes or altitudes this agency working through many years accomplishes stupendous results.

The adventurous explorer in the higher Swiss Alps frequently sees or hears the fall of fragments of rock thus loosened from the peaks.

Along the base of the vertical trap cliffs of New Haven and the Hudson River, lie immense masses of broken rock reaching to more than half the height of the bluffs themselves, rent off by this means. The same cause operates in a less conspicuous but not less important way on the surface of the stone, loosening the minute grains, as in the above instances it rends off enormous blocks. A smooth, clean pebble of the very compact Jura limestone, of such kind, for example, as abound in the rivers of South Bavaria, if moistened with water and exposed over night to sharp frost, on thawing, is muddy with the detached particles.

2.—Moving Water or Ice.

Changes of temperature not only have created differences of level in the earth's surface, but they cause a continual transfer of water from lower to higher levels. The elevated lands are cooler than the valleys. In their region occurs a continual condensation of vapor from the atmosphere, which is as continually supplied from the heated valleys. In the mountains, thus begin, as rills, the streams of water, which, gathering volume in their descent, unite below to vast rivers that flow unceasingly into the ocean.

These streams score their channels into the firmest rocks. Each grain of loosened material, as carried downward by the current, cuts the rock along which it is dragged so long as it is in motion.

The sides of the channel being undermined and loosened by exposure to the frosts, fall into the stream. In time of floods, and always, when the path of the river has a rapid descent, the mere momentum of the water acts powerfully upon any inequalities of surface that oppose its course, tearing away the rocky walls of its channel. The blocks and grains of stone, thus set in motion, grind each other to smaller fragments, and when the turbid waters clear themselves in a lake or estuary, there results a bed of gravel, sand, or soil. Two hundred and sixty years ago, the bed of the Sicilian river Simeto was obstructed by the flow across it of a stream of lava from Etna. Since that time the river, with but slight descent, has cut a channel through this hard basalt from fifty to several hundred feet wide, and in some parts forty to fifty deep.

But the action of water in pulverizing rock is not completed when it reaches the sea. The oceans are in perpetual agitation from tides, wind-waves, and currents like the Gulf-stream, and work continual changes on their shores.

Glaciers.—What happens from the rapid flow of water down the sides of mountain slopes below the frost-line is also true of the streams of ice which more slowly descend from the frozen summits. The glaciers appear like motionless ice-fields, but they are frozen rivers, rising in perpetual snows and melting into water, after having reached half a mile or a mile below the limits of frost. The snow that accumulates on the frozen peaks of high mountains, which are bathed by moist winds, descends the slopes by its own weight. The rate of descent is slow,—a few inches, or, at the most, a few feet, daily. The motion itself is not continuous, but intermittent by a succession of pushes. In the gorges, where many smaller glaciers unite, the mass has often a depth of a mile or more. Under the pressure of accumulation the snow is compacted to ice. Mingled with the snows are masses of rock broken off the higher pinnacles by the weight of adhering ice, or loosened by alternate freezing and thawing, below the line of perpetual frost. The rocks thus falling on the edge of a glacier become a part of the latter, and partake its motion. When the moving mass bends over a convex surface, it cracks vertically to a great depth. Into the *crevasses* thus formed blocks of stone fall to the bottom, and water melted from the surface in hot days flows down and finds a channel beneath the ice. The middle of the glacier moves most rapidly, the sides and bottom being retarded by friction. The ice is thus rubbed and rolled upon itself, and the stones imbedded in it crush and grind each other to smaller fragments and to dust. The rocky bed of the glacier is broken, and ploughed by the stones frozen into its sides and bottom. The glacier thus moves until it descends so low that ice cannot exist, and gradually dissolves into a torrent whose waters are always thick with mud, and whose course is strewn with worn blocks of stone (boulders) for many miles.

The Rhone, which is chiefly fed from the glaciers of the Alps, transports such a volume of rock-dust that its muddy waters may be traced for six or seven miles after they have poured themselves into the Mediterranean.

3.—CHEMICAL ACTION OF WATER AND AIR.

Water acts chemically upon rocks, or rather upon their constituent minerals, in two ways, viz., *by Combination and Solution.*

Hydration.—By chemically uniting itself to the mineral or to some ingredient of the mineral, there is formed in many instances a new compound, which, by being softer and more bulky than the original substance, is the first step towards further change. Mica, feldspar, amphibole, and pyroxene, are minerals which have been artificially produced in the slags or linings of smelting furnaces, and thus formed they have been found totally destitute of water, as might be expected from the high temperature in which they originated. Yet these minerals as occurring in nature, even when broken out of blocks of apparently unaltered rock, and especially when they have been directly exposed to the weather, often, if not always, contain a small amount of water, in chemical combination (water of hydration).

Solution.—As a solvent, water exercises the most important influence in disintegrating minerals. Apatite, when containing much chlorine, is gradually decomposed by treatment with water, chloride of calcium, which is very soluble, being separated from the nearly insoluble phosphate of lime. The minerals which compose silicious rocks are all acted on perceptibly by pure water. This is readily observed when the minerals are employed in the state of fine powder. If pulverized feldspar, amphibole, etc., are simply moistened with pure water, the latter at once dissolves a trace of alkali, as shown by its turning red litmus-paper blue. This solvent action is so slight upon a smooth mass of the mineral as hardly to be perceptible, because the action is limited by the extent of surface. Pulverization, which increases the surface enormously, increases the solvent effect in a similar proportion. A glass vessel may have water boiled in it for hours without its luster being dimmed or its surface materially acted upon, whereas the same glass* finely pulverized is attacked by water so readily as to give at once a solution alkaline to the taste. Messrs. W. B. and R. E. Rogers (*Am. Jour. Sci.*, V, 404, 1848) found that by continued digestion of pure water for a week, with powdered feldspar, hornblende, chlorite, serpentine, and natrolite,† these minerals yielded to the solvent from 0.4 to 1 per cent of their weight.

In nature we never deal with pure water, but with water holding in solution various matters, either derived from the air or from the soil. These substances modify, and in most cases enhance, the solvent power of water.

Action of Carbonic Acid.—This gaseous substance is absorbed by or dissolved in all natural waters to a greater or less extent. At common temperatures and pressure water is capable of taking up its own bulk of the gas. At lower temperatures, and under increased pressure, the quantity dissolved is much greater. *Carbonated water*, as we may designate this solution, has a high solvent power on the carbonates of lime, magnesia, protoxide of iron, and protoxide of manganese. The salts just named are as good as insoluble in pure water, but they exist in considerable quantities in most natural waters. The spring and well waters of limestone regions are *hard* on account of their content of carbonate of lime. *Chalybeate waters* are those which hold carbonate of iron in solution. When carbonated water comes in contact with silicious minerals, these are decomposed much more rapidly than by pure water. The lime, magnesia, and iron they contain, are partially removed in the form of carbonates.

Struve exposed powdered phonolite (a rock composed of feldspar and zeolites) to water saturated with carbonic

---

* Glass is a silicate of potash or soda.

† Mesotype.

acid under a pressure of 3 atmospheres, and obtained a solution of which a pound* contained:

Carbonate of soda,	22.0 grains.
Chloride of sodium,	2.0 "
Sulphate of potash,	1.7 "
" " soda,	4.8 "
Carbonate of lime,	4.5 "
" " magnesia,	1.1 "
Silica,	0.5 "
Phosphoric acid and manganese, traces	
Total,	37.1 grains.

In various natural springs, water comes to the surface so charged with carbonic acid that the latter escapes copiously in bubbles. Such waters dissolve large quantities of mineral matters from the rocks through which they emerge. Examples are seen in the springs at Saratoga, N. Y. According to Prof. Chandler, the "Saratoga Spring," whose waters issue directly from the rock, contains in one gallon of 231 cubic inches:

Chloride of Sodium (common salt)	398.361	grains.
" " Potassium,	9.698	"
Bromide of Sodium,	0.571	"
Iodide of Sodium,	0.126	"
Sulphate of Potash,	5.400	"
Carbonate of Lime,	86.483	"
" " Magnesia,	41.050	"
" " Soda,	8.948	"
" " Protoxide of iron,	.879	"
Silica,	1.283	"
Phosphate of lime,	trace	"
Solid matters,	552.799	"
Carbonic acid gas, (407.647 cubic inches at 52° Fah.)		
Water,	58,317.110	"

The waters of ordinary springs and rivers, as well as those that fall upon the earth's surface as rain, are, indeed,

---

* The Saxon pound contains 7,680 Saxon grains.

by no means fully charged with carbonic acid, and their solvent effect is much less than that exerted by water saturated with this gas.

The quantity (by volume) of carbonic acid in 10,000 parts of rain-water has been observed as follows: According to

		Locality.
Lampadius,	8	Country near Freiberg, Saxony.
Mulder,	20	City of Utrecht, Holland.
Von Baumhauer,	40 to 90	" " " "
Pellgot,	5	?

The quantities found are variable, as might be expected, and we notice that the largest proportion above cited does not even amount to one per cent.

In river and spring water the quantities are somewhat larger, but the carbonic acid exists chiefly in chemical combination as bicarbonates of lime, magnesia, etc.

In the capillary water of soils containing much organic matters, more carbonic acid is dissolved. According to a single observation of De Saussure's, such water contains $2°|_0$ of the gas. In a subsequent paragraph, p. 221, is given the reason of the small content of carbonic acid in these waters.

The weaker action of these dilute solutions, when continued through long periods of time and extending over an immense surface, nevertheless accomplishes results of vast significance.

Solutions of Alkali-Salts.—Rain-water, as we have already seen, contains a minute quantity of salts of ammonia (nitrate and bicarbonate). The water of springs and rivers acquires from the rocks and soil, salts of soda and potash, of lime and magnesia. These solutions, dilute though they are, greatly surpass pure water, or even carbonated water, in their solvent and disintegrating action. Phosphate of lime, the earth of bones, is dissolved by pure water to an extent that is hardly appreciable; in salts of ammonia and of soda, however, it is taken up in considerable quantity. Solution of nitrate of ammonia dissolves lime and magnesia and their carbonates with great ease. In general, up to a certain limit, a saline solution acquires increased solvent power by increase in the amount and number of dissolved matters. This important fact is one to which we shall recur at another time.

Action of Oxygen.—This element, the great mover of chemical changes, which is present so largely in the atmosphere, has a strong tendency to unite with certain bodies which are almost universally distributed in the rocks. On turning to the analyses of minerals, p. 110, we notice in nearly every instance a quantity of protoxide of iron, or protoxide of manganese. The green, dark gray, or black minerals, as the micas, amphibole, pyroxene, chlorite, talc, and serpentine, invariably contain these protoxides in notable proportion. In the feldspars they exist, indeed, in very minute quantity, but are almost never entirely wanting. Sulphide of iron (iron pyrites), in many of its forms, is also disposed to oxidize its sulphur to sulphuric acid, its iron to sesquioxide, and this mineral is widely distributed as an admixture in many rocks. In trap or basaltic rocks, as at Bergen Hill, *metallic iron* is said to occur in minute proportion,* and in a state of fine division. The oxidation of these substances materially hastens the disintegration of the rocks containing them, since the higher oxides of iron and of manganese occupy more space than the metals or lower oxides. This fact is well illustrated by the sulphate of protoxide of iron (copperas, or green-vitriol), which, on long keeping, exposed to the air, is converted from transparent, glassy, green crystals to a bulky, brown, opaque powder of sulphate of sesquioxide of iron.

Weathering.—The conjoined influence of water, car-

---

* This statement rests on the authority of Professor Henry Wurtz, of New York.

bonic acid, oxygen, and the salts held in solution by the atmospheric waters, is expressed by the word *weathering*. This term may likewise include the action of frost.

When rocks weather, they are decomposed or dissolved, and new compounds, or new forms of the original matter, result. The soil is a mixture of broken or pulverized rocks, with the products of their alteration by weathering.

a. **Weathering of Quartz Rock.**—Quartz (silicic acid), as occurring nearly pure in quartzite, and in many sandstones, or as a chief ingredient of all the granitic, hornblendic, and many other rocks, is so exceedingly hard and insoluble, that the lifetime of a man is not sufficient for the direct observation of any change in it, when it is exposed to ordinary weathering. It is, in fact, the least destructible of the mineral elements of the globe. Nevertheless, quartz, even when pure, is not absolutely insoluble, particularly in water containing alkali carbonates or silicates. In its less pure varieties, and especially when associated with readily decomposable minerals, it is acted on more rapidly. The quartz of granitic rocks is usually roughened on the surface when it has long been exposed to the weather.

b. **The Feldspars** weather much more easily than quartz, though there are great differences among them. The soda and lime feldspars decompose most readily, while the potash feldspars are often exceedingly durable. The decomposition results in completely breaking up the hard, glassy mineral. In its place there remains a white or yellowish mass, which is so soft as to admit of crushing between the fingers, and which, though usually, to the naked eye, opaque, and non-crystalline, is often seen, under a powerful magnifier, to contain numerous transparent crystalline plates. The mass consists principally of the crystalline mineral, *kaolinite*, a hydrated silicate of alumina, (the analysis of which has been given already, p. 113,) mixed with hydrated silica, and often with grains of undecomposed mineral. If we compare the composition of pure potash feldspar with that of kaolinite, assuming, what is probably true, that all the alumina of the former remains in the latter, we find what portions of the feldspar have been removed and washed away by the water, which, together with carbonic acid, is the agent of this change.

	Feldspar.	Kaolinite.	Liberated.	Added.
Alumina	18.3	18.3	0	
Silica	64.8	23.0	41.8	
Potash	16.9		16.9	
Water		6.4		6.4
	100	47.7	58.7	6.4

It thus appears that, in the complete conversion of 100 parts of potash feldspar into kaolinite, there result 47.7 parts of the latter, while $58.7°|_0$ of the feldspar, viz: $41.8°|_0$ of silica and $16.9°|_0$ of potash, are dissolved out.

The potash, and, in case of other feldspars, soda, lime, and magnesia, are dissolved as carbonates. If much water has access during the decomposition, all the liberated silica is carried away.* It usually happens, however, that a portion of the silica is retained in the kaolin (perhaps in a manner similar to that in which bone charcoal retains the coloring matters of crude sugar). The same is true of a portion of the alkali, lime, and oxide of iron, which may have existed in the original feldspar.

The formation of kaolin may be often observed in nature. In mines, excavated in feldspathic rocks, the fissures and cavities through which surface water finds its way downwards are often coated or filled with this substance.

c. **Other Silicious Minerals,** as Leucite, (Topaz, Scapolite,) etc., yield kaolin by decomposition. It is probable that the micas, which decompose with difficulty, (phlogo-

---

* We have seen (H. C. G., p. 121) that silica, when newly set free from combination, is, at first, freely soluble in water.

pite, perhaps, excepted,) and the amphiboles and pyroxenes, which are often easily disintegrated, also yield kaolin; but in the case of these latter minerals, the resulting kaolinite is largely mixed with oxides and silicates of iron and manganese, so that its properties are modified, and identification is difficult. Other hydrated silicates of alumina, closely allied to kaolinite, appear to be formed in the decomposition of compound silicates.

Ordinary Clays, as pipe-clay, blue-clay, brick-clay, etc., are mixtures of kaolinite, or of a similar hydrated silicate of alumina, with a variety of other substances, as free silica, oxides, and silicates of iron and manganese, carbonate of lime, and fragments or fine powder of undecomposed minerals. Fresenius deduces from his analyses of several Nassau clays the existence in them of a compound having the symbol $Al_2 O_3\ 3\ SiO_2 + H_2O$, and the following composition *per cent.*

Silica,	57.14
Alumina,	31.72
Water,	11.14
	100.00

Other chemists have assumed the existence of hydrated silicates of alumina of still different composition in clays, but kaolinite is the only one which occurs in a pure state, as indicated by its crystallization, and the existence of the others is not perfectly established. (S. W. Johnson and J. M. Blake *on Kaolinite, etc., Am. Jour. Sci., May,* 1867, *pp.* 351–362.)

d. **The Zeolites** readily suffer change by weathering; little is known, however, as to the details of their disintegration. Instead of yielding kaolinite, they appear to be transformed into other zeolites, or retain something of their original chemical constitution, although mechanically disintegrated or dissolved. We shall see hereafter that there is strong reason to assume the existence of compounds analogous to zeolites in every soil.

e. **Serpentine** and **Magnesian Silicates** are generally slow of decomposition, and yield a meager soil.

f. The **Limestones**, when pure and compact, are very durable: as they become broken, or when impure, they often yield rapidly to the weather, and impregnate the streams which flow over them with carbonate of lime.

g. **Argillite** and **Argillaceous Limestones**, which have resulted from the solidification of clays, readily yield clay again, either by simple pulverization or by pulverization and weathering, according as they have suffered more or less change by metamorphism.

§ 5.

INCORPORATION OF ORGANIC MATTER WITH THE SOIL AND ITS EFFECTS.

Antiquity of Vegetation.—Geological observations lead to the conclusion that but small portions of the earth's surface-rocks were formed previous to the existence of vegetation. The enormous tracts of coal found in every quarter of the globe are but the residues of preadamite forests, while in the oldest stratified rocks the remains of plants (marine) are either most distinctly traced, or the abundance of animal forms warrants us in assuming the existence of vegetation previous to their deposition.

The Development of Vegetation on a purely Mineral Soil.—The mode in which the original inorganic soil became more or less impregnated with organic matter may be illustrated by what has happened in recent years upon the streams of lava that have issued from volcanoes. The lava flows from the crater as red-hot molten rock, often in masses of such depth and extent as to require months to cool down to the ordinary temperature. For many years the lava is incapable of bearing any vegetation save some almost microscopic forms. During these years the surface of the rock suffers gradual disintegration by the agencies of air and water, and so in time acquires the power to support some lichens that appear at first as mere stains upon its surface. These, by their decay, increase the film of soil from which they sprung. The growth of new generations of these plants is more and more vigorous, and other superior kinds take root among them. After another period of years, there has accumulated a tangible soil, supporting herbaceous plants and dwarf shrubs. Henceforward the increase proceeds more rapidly; shrubs gradually give place to trees, and in a century, more or less, the once hard, barren rock has weathered to a soil fit for vineyards and gardens.

Those lowest orders of plants, the lichens and mosses, which prepare the way for forests and for agricultural vegetation, are able to extract nourishment from the most various and the most insoluble rocks. They occur abundantly on all our granitic and schistose rocks. Even on quartz they do not refuse to grow. The white quartz hills of Berkshire, Massachusetts, are covered on their moister northern slopes with large patches of a leathery lichen, which adheres so firmly to the rock that, on being forced off, particles of the stone itself are detached. Many of the old marbles of Greece are incrusted with oxalate of lime left by the decay of lichens which have grown upon their surface.

Humus.—By the decay of successive generations of plants the soil gradually acquires a certain content of dead organic matter. The falling leaves, seeds and stems of vegetation do not in general waste from the surface as rapidly as they are renewed. In forests, pastures, prairies, and marshes, there accumulates on the surface a brown or black mass, termed *humus*, of which leaf-mold, swamp-muck, and peat are varieties, differing in appearance as in the circumstances of their origin. In the depths of the soil similar matters are formed by the decay of roots and other subterranean parts of plants, or by the inversion of sod and stubble, as well as by manuring.

Decay of Vegetation.—When a plant or any part of a plant dies, and remains exposed to air and moisture at the common temperatures, it undergoes a series of chemical and physical changes, which are largely due to an oxidation of portions of its carbon and hydrogen, and the formation of new organic compounds. Vegetable matter is considerably variable in composition, but in all cases chiefly consists of cellulose and starch, or bodies of similar character, mixed with a small proportion of albuminous and mineral substances. By decay, the white or light-colored and tough tissues of plants become converted into brown or black friable substances, in which less or none of the organized structure of the fresh plant can be traced. The bulk and weight of the decaying matter constantly decreases as the process continues. With full access of air and at suitable temperatures, the decay, which, from the first, is characterized by the production and escape of carbonic acid and water, proceeds without interruption, though more and more slowly, until nearly all the carbon and hydrogen of the vegetable matters are oxidized to the above-named products, and little more than the ashes of the plant remains. With limited access of air the process rapidly runs through a first stage of oxidation, when it becomes checked by the formation of substances which are themselves able, to a good degree, to resist further oxidation, especially under the circumstances of their formation, and hence they accumulate in considerable quantities. This happens in the lower layers of fallen leaves in a dense forest, in compost and manure heaps, in the sod of a meadow or pasture, and especially in swamps and peat-bogs.

The more delicate, porous and watery the vegetable

matter, and the more soluble substances and albuminoids it contains, the more rapidly does it decay or humify.

It has been shown by a chemical examination of what escapes in the form of gas, as well as of what remains as humus, that the carbon of wood oxidizes more slowly than its hydrogen, so that humus is relatively richer in carbon than the vegetable matters from which it originates. With imperfect access of air, carbon and hydrogen are to some extent disengaged in union with each other, as marsh gas (CH_4). Carbonic oxide gas (CO) is probably also produced in minute quantity. The nitrogen of the vegetable matter is to a considerable extent liberated in the free gaseous state; a portion of it unites to hydrogen, forming ammonia (NH_3), which remains in the decaying mass; still another portion remains in the humus in combination, not as ammonia, but as an ingredient of the ill-defined acid bodies which constitute the bulk of humus; finally, some of the nitrogen may be oxidized to nitric acid.

Chemical Nature of Humus.—In a subsequent chapter, (p. 224,) the composition of humus will be explained at length. Here we may simply mention that, under the influence of alkalies and ammonia, it yields one or more bodies having acid characters, called humic and ulmic (also geic) acids. Further, by oxidation it gives rise to crenic and apocrenic acids. The former are faintly acid in their properties; the latter are more distinctly characterized acids.

Influence of Humus on the Minerals of the Soil.—
a. Disintegration of the mineral matters of soils is aided by the presence of organic substances in a decaying state, in so far as the latter, from their hygroscopic quality, maintain the surface of the soil in a constant state of *moisture.*

b. Organic matters furnish copious supplies of *carbonic acid,* the action of which has already been considered

(p. 128). Boussingault and Lewy (*Mémoires de Chimie Agricole, etc., p.* 369,) have analyzed the air contained in the pores of the soil, and, as was to be anticipated, found it vastly richer in carbonic acid than the ordinary atmosphere.

The following table exhibits the composition of the air in the soil compared with that of the air above the soil, as observed in their investigations.

				Carbonic acid in 10,000 parts of air (by weight).
Ordinary atmosphere				6
Air from sandy subsoil of forest				38
"	"	loamy	" "	124
"	"	surface-soil	" "	130
"	"	"	" vineyard	146
"	"	"	" old asparagus bed	122
"	"	"	" " " newly manured	233
"	"	"	" pasture	270
"	"	"	rich in humus	543
"	"	"	newly manured sandy field, during dry weather	333
"	"	"	newly manured sandy field, during wet weather	1413

That this carbonic acid originates in large part by oxidation of organic matters is strikingly demonstrated by the increase in its quantity, resulting from the application of manure, and the supervention of warm, wet weather. It is obvious that the carbonic acid contained in the air of the soil, being from twenty to one hundred or more times more abundant, relatively, than in the common atmosphere, must act in a correspondingly more rapid and energetic manner in accomplishing the solution and disintegration of mineral matters.

c. The *organic acids of the humus group* probably aid in the disintegration of soil by direct action, though our knowledge is too imperfect to warrant a positive conclusion. The ulmic and humic acids themselves, indeed, do not, according to Mulder, exist in the free state in the soil, but their soluble salts of ammonia, potash or soda, have acid characters, in so far that they unite energetically with other bases, as lime, oxide of iron, etc. These alkali-salts, then, should attack the minerals of the soil in a manner similar to carbonic acid. The same is probably true of crenic and apocrenic acids.

d. It scarcely requires mention that the *ammonia salts* and *nitrates* yielded by the decay of plants, as well as the organic acids, oxalic, tartaric, etc., or acid-salts, and the chlorides, sulphates, and phosphates they contain, act upon the surface soil where they accumulate in the manner already described, and that vegetable (and animal) remains thus indirectly hasten the solution of mineral matters.

Action of Living Plants on the Minerals of the Soil.—

1. *Moisture and Carbonic Acid.*—The living vegetation of a forest or prairie is the means of perpetually bringing the most vigorous disintegrating agencies to bear upon the soil that sustains it. The shelter of the growing plants, not less than the hygroscopic humus left by their decay, maintains the surface in a state of saturation by moisture. The carbonic acid produced in living roots, and to some extent, at least, it is certain, excreted from them, adds its effect to that derived from other sources.

2. *Organic Acids within the Plant.*—According to Zöller, (*Vs. St.* V. 45) the young roots of living plants (what plants, is not mentioned) contain an acid or acid-salt which so impregnates the tissues as to manifest a strong acid reaction with (give a red color to) blue litmus-paper, which is permanent, and therefore not due to carbonic acid. This acidity, Zöller informs us, is most intense in the finest fibrils, and is exhibited when the roots are simply wrapped in the litmus-paper, without being at all (?) crushed or broken. The acid, whatever it may be, thus existing within the roots is absorbed by porous paper placed externally to them.

Previous to these observations of Zöller, Salm Horstmar (*Jour. für. Prakt. Chem.* XL. 304,) having found in the ashes of ground pine (*Lycopodium complanatum*), $38°|_0$ of alumina, while in the ashes of juniper, growing beside the Lycopodium, this substance was absent, examined the rootlets of both plants, and found that the former had an acid reaction, while the latter did not affect litmus-paper. Salm Horstmar supposed that the alumina of the soil finds its way into the Lycopodium by means of this acid. Ritthausen has shown that the Lycopodium contains malic acid, and since all the alumina of the plant may be extracted by water, it is probable that the acid reaction of the rootlets is due, in part at least, to the presence of acid malate of alumina. (*Jour. für. Prakt. Chem*. LIII. 420.)

At Liebig's suggestion, Zöller made the following experiments. A number of glass tubes were filled with water made slightly acid by some drops of hydrochloric acid, vinegar, citric acid, bitartrate of potash, etc.; the open end of each tube was then closed by a piece of moistened bladder tied tightly over, and various salts, insoluble in water, as phosphate of lime, phosphate of ammonia and magnesia, etc., were strewn on the bladder. After a short time it was found that the ingredients of these salts were contained in the liquid in contact with the under surface of the bladder, having been dissolved by the dilute acid present in the pores of the membrane, and absorbed through it. This is an ingenious illustration of the mode in which the organic acids existing in the root-cells of plants may act directly upon the rock or soil external to them. By such action is doubtless to be explained the fact mentioned by Liebig in the following words:

"We frequently find in meadows smooth limestones with their surfaces covered with a network of small furrows. When these stones are newly taken out of the ground, we find that each furrow corresponds to a rootlet, which appears as if it had eaten its way into the stone." (*Modern Ag* p. 43.)

This direct action of the living plant is probably exerted by the lichens, which, as has been already stated, grow upon the smooth surface of the rock itself. Many of the lichens are known to contain oxalate of lime to the extent of half their weight (Braconnot).

According to Goeppert, the hard, fine-grained rock of the Zobtenberg, a mountain of Silesia, is in all cases softened at its surface where covered with lichens (*Acarospora smaragdula*, *Imbricaria olivacea*, etc.), while the bare rock, closely adjacent, is so hard as to resist the knife. On the Schwalbenstein, near Glatz, in Silesia, at a height of 4,500 feet, the granite is disintegrated under a covering of lichens, the feldspar being converted into kaolin or washed away, only the grains of quartz and mica remaining unaltered.*

CHAPTER III.

KINDS OF SOILS—THEIR DEFINITION AND CLASSIFICATION.

§ 1.

DISTINCTION OF SOILS BASED UPON THE MODE OF THEIR FORMATION OR DEPOSITION.

The foregoing considerations of the origin of soils introduce us appropriately to the study of soils themselves. In the next place we may profitably recount those definitions and distinctions that serve to give a certain degree of precision to language, and enable us to discriminate in some measure the different kinds of soils, which offer great diversity in origin, composition, external characters,

---

* See, also, p. 136.

and fertility. Unfortunately, while there are almost numberless varieties of soil having numberless grades of productive power, we are very deficient in terms by which to express concisely even the fact of their differences, not to mention our inability to define these differences with accuracy, or our ignorance of the precise nature of their peculiarities.

As regards mode of formation or deposition, soils are distinguished into *Sedentary* and *Transported*. The latter are subdivided into *Drift*, *Alluvial*, and *Colluvial* soils.

Sedentary Soils, or *Soils in place*, are those which have not been transported by geological agencies, but which remain where they were formed, covering or contiguous to the rock from whose disintegration they originated. Sedentary soils have usually little depth. An inspection of the rock underlying such soils often furnishes most valuable information regarding their composition and probable agricultural value; because the still unweathered rock reveals to the practised eye the nature of the minerals, and thus of the elements, composing it, while in the soil these may be indistinguishable.

In New England and the region lying north of the Ohio and east of the Missouri rivers, soils in place are not abundant as compared with the entire area. Nevertheless they do occur in many small patches. Thus the red-sandstone of the Connecticut Valley often crops out in that part of New England, and, being, in many localities, of a friable nature, has crumbled to soil, which now lies undisturbed in its original position. So, too, at the base of trap-bluffs may be found trap-soils, still full of sharp-angled fragments of the rock.

Transported Soils, (subdivided into drift, alluvial, and colluvial), are those which have been removed to a distance from the rock-beds whence they originated, by the action of moving ice (glaciers) or water (rivers), and deposited as *sediment* in their present positions.

Drift Soils (sometimes called diluvial) are characterized by the following particulars. They consist of fragments whose edges at least have been rounded by friction, if the fragments themselves are not altogether destitute of angles. They are usually deposited without any stratification or separation of parts. The materials consist of soil proper, mingled with stones of all sizes, from sand-grains up to immense rock-masses of many tons in weight. This kind of soil is usually distinguished from all others by the rounded rocks or *boulders* ("hard heads") it contains, which are promiscuously scattered through it.

The "Drift" has undoubtedly been formed by moving ice in that period of the earth's history known to geologists as the Glacial Epoch, a period when the present surface of the country was covered to a great depth by fields of ice.

In regions like Greenland and the Swiss Alps, which reach above the line of perpetual snow, drift is now accumulating, perfectly similar in character to that of New England, or has been obviously produced by the melting of glaciers, which, in former geological ages and under a colder climate, were continuations on an immense scale of those now in existence.

A large share of the northern portion of the country from the Arctic regions southward as far as latitude 39°, or nearly to the southern boundaries of Pennsylvania and to the Ohio River, including Canada, New England, Long Island, and the States west as far as Iowa, is more or less covered with drift. Comparison of the boulders with the undisturbed rocks of the regions about show that the materials of the drift have been moved southwards or southeastwards to a distance generally of twenty to forty miles, but sometimes also of sixty or one hundred miles, from where they were detached from their original beds.

The surface of the country when covered with drift is often or usually irregular and hilly, the hills themselves being conical heaps or long ridges of mingled sand, gravel, and boulders, the transported mass having often a great depth. These hills or ridges are parts of the vast trains of material left by the melting of preadamite glaciers or icebergs, and have their precise counterpart in the *moraines* of the Swiss Alps. Drift is accordingly not confined to the valleys, but the northern slopes of mountains or hills, whose basis is unbroken rock, are strewn to the summit with it, and immense blocks of transported stone are seen upon the very tops of the Catskills and of the White and Green Mountains.

Drift soils are for these reasons often made up of the most diverse materials, including all the kinds of rock and rock-dust that are to be found, or have existed for one or several scores of miles to the northward. Of these often only the harder granitic or silicious rocks remain in considerable fragments, the softer rocks having been completely ground to powder.

Towards the southern limit of the Drift Region the drift itself consists of fine materials which were carried on by the waters from the melting glaciers, while the heavier boulders were left further north. Here, too, may often be observed a partial stratification of the transported materials as the result of their deposition from moving water. The great belts of yellow and red sand that stretch across New Jersey on its southeastern face, and the sands of Long Island, are these finer portions of the drift. Farther to the north, many large areas of sand may, perhaps, prove on careful examination to mark the southern limit of some ancient local glacier.

Alluvial Soils consist of worn and rounded materials which have been transported by the agency of running water (rivers and tides). Since small and light particles are more readily sustained in a current of water than

7

heavy masses, alluvium is always more or less *stratified* or arranged in distinct layers: stones or gravel at the bottom and nearest the source of movement, finer stones or finer gravel above and further down in the path of flow, sand and impalpable matters at the surface and at the point where the stream, before turbid from suspended rock-dust, finally clears itself by a broad level course and slow progress.

Alluvial deposits have been formed in all periods of the earth's history. Water trickling gently down a granite slope carries forward the kaolinite arising from decomposition of feldspar, and the first hollow gradually fills up with a bed of clay. In valleys are thus deposited the gravel, sand, and rock-dust detached from the slopes of neighboring mountains. Lakes and gulfs become filled with silt brought into them by streams. Alluvium is found below as well as above the drift, and recent alluvium in the drift region is very often composed of drift materials rearranged by water-currents. Alluvium often contains rounded fragments or disks of soft rocks, as limestones and slates, which are more rarely found in drift.

Colluvial Soils, lastly, are those which, while consisting in part of drift or alluvium, also contain sharp, angular fragments of the rock from which they mainly originated, thus demonstrating that they have not been transported to any great distance, or are made up of soils in place, more or less mingled with drift or alluvium.

§ 2.

DISTINCTIONS OF SOILS BASED UPON OBVIOUS OR EXTERNAL CHARACTERS.

The classification and nomenclature of soils customarily employed by agriculturists have chiefly arisen from consideration of the relative proportions of the principal mechanical ingredients, or from other highly obvious qualities.

The distinctions thus established, though very vague scientifically considered, are extremely useful for practical purposes, and the grounds upon which they rest deserve to be carefully reviewed for the purpose of appreciating their deficiencies and giving greater precision to the terms employed to define them.

The farmer, speaking of soils, defines them as *gravelly, sandy, clayey, loamy, calcareous, peaty, ochreous,* etc.

Mechanical Analysis of the Soil.—Before noticing these various distinctions in detail, we may appropriately study the methods which are employed for separating the mechanical ingredients of a soil. It is evident that the epithet *sandy*, for example, should not be applied to a soil unless sand be the predominating ingredient; and in order to apply the term with strict correctness, as well as to know how a soil is constituted as regards its mechanical elements, it is necessary to isolate its parts and determine their relative quantity.

Boulders, stones, and pebbles, are of little present or immediate value in the soil by way of feeding the plant. This function is performed by the finer and especially by the finest particles. Mechanical analysis serves therefore to compare together different soils, and to give useful indications of fertility. Simple inspection aided by the feel enables one to judge, perhaps, with sufficient accuracy for all ordinary practical purposes; but in any serious attempt to define a soil precisely, for the purposes of science, its mechanical analysis must be made with care.

Mechanical separation is effected by *sifting* and *washing*. Sifting serves only to remove the *stones* and *coarse sand*. By placing the soil in a glass cylinder, adding water, and vigorously agitating for a few moments, then letting the whole come to rest, there remains suspended in the water a greater or less quantity of matter in a state of extreme division. This fine matter is in many cases clay (kaolinite), or at least consists of substances resulting from the weathering of the rocks, and is not, or not chiefly, rock-dust. Between this impalpably fine matter and the grains of sand retained by a sieve, there exist numberless gradations of fineness in the particles.

By conducting a slow stream of water through a tube to the bottom of a vessel, the fine particles of soil are carried off and may be received in a pan placed beneath. Increasing the rapidity of the current enables it to remove larger particles, and thus it is easy to separate the soil into a number of portions, each of which contains soil of a different fineness.

Various attempts have been made to devise precise means of separating the materials of soils mechanically into a definite number of grades of fineness.

This may be accomplished in good measure by washing, but constant and accurate results are of course only attained when the circumstances of the washing are uniform throughout. The method adopted by the Society of Agricultural Chemists of Germany is essentially the following (*Versuchs Stationen*, VI, 144):

The *air-dry* soil is gently rubbed on a tin-plate sieve with round holes three millimeters in diameter; what passes is weighed as *fine-earth*. What remains on the sieve is washed with water, dried, weighed, and designated as gravel, pebbles, stones, as the case may be, the size of the stones, etc., being indicated by comparison with the fist, with an egg, a walnut, a hazelnut, a pea, etc. Of the *fine-earth* a portion (30 grams) is now boiled for an hour or more in water, so as to completely break down any lumps and separate adhering particles, and is then left at rest for some minutes, when it is transferred into the vessel 1 of the apparatus, fig. 8., after having poured off the turbid water with which it was boiled, into 2. This washing apparatus (invented by Nöbel) consists of a reservoir, *A*, made of sheet metal, capable of holding something more than 9 liters of water, and furnished at *b* with a stop-cock. By means of a tube of rubber it is joined to the series of

Fig. 8.

vessels, **1, 2, 3,** and **4,** which are connected to each other, as shown in the figure, the recurved neck of **2** fitting water-tight into the nozzle of **1** at *a*, etc.

These vessels are made of glass, and together hold 4 liters of water; their relative volume is nearly

$$1 : 8 : 27 : 64, \text{ or } = 1^3 : 2^3 : 3^3 : 4^3.$$

5 is a glass vessel of somewhat more than 5 liters capacity.

The distance between *b* and *c* is 2 feet. The cock, *b*, is opened, so that in 20 minutes exactly 9 liters of water pass it. The apparatus being joined together, and the cock opened, the soil in 1 is agitated by the stream of water flowing through, and the finer portions are carried over into 2, 3, 4, and 5. As a given amount of water requires eight times longer to pass through 2 than 1, its velocity of motion and buoyant power in the neck of 3 are correspondingly less. After the requisite amount of water has run from *A*, the cock is closed, the whole left to rest several hours, when the contents of the vessels are separately rinsed out into porcelain dishes, dried and weighed.*

The contents of the several vessels are designated as follows:†

1. Gravel, fragments of rock.
2. Coarse sand.
3. Fine sand.
4. Finest or dust sand.
5. Clayey substance or impalpable matter.

In most inferior soils the *gravel*, the *coarse sand*, and the *fine sand*, are angular fragments of quartz, feldspar, amphibole, pyroxene, and mica, or of rocks consisting of these minerals. It is only these harder and less easily decomposable minerals that can resist the pulverizing agencies through which a large share of our soils have passed. In the more fertile soils, formed from sedimentary limestones and slates, the fragments of these stratified rocks occur as flat pebbles and rounded grains.

The *finest* or *dust-sand*, when viewed under the microscope, is found to be the same rocks in a higher state of pulverization.

---

* See, also, Wolff's "*Anleitung zur Untersuchung landwirthschaftlich-wichtiger Stoffe,*" 1867, p. 5.

† These names, applied by Wolff to the results of washing the sedentary soils of Würtemberg, do not always well apply to other soils. Thus Grouven, (3*ter Salzmünder Bericht*, p. 32), operating on the alluvial soils of North Germany, designated the contents of the 4th funnel as "clay and loam," and those of the 5th vessel as "light clay and humus." Again, Schöne found (*Bulletin, etc., de Moscou*, p. 402) by treatment of a certain soil in Nöbel's apparatus, 45 per cent of "coarse sand" remaining in the 2d funnel. The particles of this were for the most part smaller than 1-10th millimeter (1-250th inch), which certainly is not coarse sand!

What is designated as *clayey substance*, or *impalpable matter*, is oftentimes largely made up of rock-dust, so fine that it is supported by water, when the latter is in the gentlest motion. In what are properly termed clay-soils, the finest parts consist, however, chiefly of the hydrous silicate of alumina, already described, p. 113, under the mineralogical name of *kaolinite*, or of analogous compounds, mixed with gelatinous silica, oxides of iron, and carbonate of lime, as well as with finely divided quartz and other granitic minerals. So gradual is the transition from true kaolinite clay through its impurer sorts to mere impalpable rock-dust, in all that relates to sensible characters, as color, feel, adhesiveness, and plasticity, that the term clay is employed rather loosely in agriculture, being not infrequently given to soils that contain very little kaolinite or true clay, and thus implies the general physical qualities that are usually typified by clay rather than the presence of any definite chemical compound, like kaolinite, in the soil.

Many soils contain much *carbonate of lime* in an impalpable form, this substance having been derived from lime rocks, as marble and chalk, from the shells of mollusks, or from coral; or from clays that have originated by the chemical decomposition of feldspathic rocks containing much lime.

Organic matter, especially the *debris* of former vegetation, is almost never absent from the impalpable portion of the soil, existing there in some of the various forms assumed by humus.

As Schöne has shown, (*Bulletin de la Societé des Naturalistes de Moscou*, 1867, p. 363), the results obtained by Nöbel's apparatus are far from answering the purposes of science. The separation is not carried far enough, and no simple relations subsist between the separated portions, as regards the dimensions of their particles. If the soil were composed of spherical particles of one kind of matter, or having all the same specific gravity, it would be possible by the use of a properly constructed washing apparatus to separate a sample into fifty or one hundred parts, and to define the dimensions of the particles of each of these parts. Since, however, the soil is very heterogeneous, and since its particles are unlike in shape, consisting partly of nearly spherical grains and partly of plates or scales upon which moving water exerts an unequal floating effect, it is difficult, if not impossible, to realize so perfect a mechanical analysis. It is, however, easy to make a separation of a soil into a large number of parts, each of which shall admit of precise definition in terms of the rapidity of flow of a current of water capable of sustaining the particles which compose it. Instruments for mechanical analysis, which provide for producing and maintaining at will any desired rate of flow in a stream of water, have been very recently devised, independently of each other, by E. Schöne (*loc. cit.*, pp. 334-405) and A. Muller (*Vs. St.*, X, 25-51). The employment of such apparatus promises valuable results, although as yet no extended investigations made with its help have been published.

Gravelly Soils are so named from the abundance of small stones or pebbles in them. This name alone gives but little idea of the really important characters of the soil. Simple gravel is nearly valueless for agricultural purposes; many highly gravelly soils are, however, very fertile. The fine portion of the soil gives them their crop-feeding power. The coarse parts ensure drainage and store the solar heat. The mineralogical characters of the pebbles in a soil, as determined by a practised eye, may often give useful indications of its composition, since it is generally true that the finer parts of the soil agree in this respect with the coarser, or, if different, are not inferior. Thus if the gravel of a soil contains many pebbles of feldspar, the soil itself may be concluded to be well supplied with alkalies; if the gravel consists of limestone, we may infer that lime is abundant in the soil. On the other hand, if a soil contains a large proportion of quartz pebbles, the legitimate inference is that it is of comparatively poor quality. The term gravelly admits of various qualification. We may have a very gravelly or a moderately gravelly soil, and the coarse material may be characterized as a fine or coarse gravel, a slaty gravel, a granitic gravel, or a diorite gravel, according to its state of division or the character of the rock from which it was formed.

But the closest description that can thus be given of a gravelly soil cannot convey a very precise notion of even its external qualities, much less of those properties upon which its fertility depends.

Sandy Soils are those which visibly consist to a large degree, $90°|_0$ or more, of *sand*, i. e., of small granular fragments of rock, no matter of what kind. Sand usually signifies grains of *quartz*; this mineral, from its hardness, withstanding the action of disintegrating agencies beyond any other. Considerable tracts of nearly pure and white quartz sand are not uncommon, and are characterized by obdurate barrenness. But in general, sandy soils are by no means free from other silicious minerals, especially feldspar and mica. When the sand is yellow or red in color, this fact is due to admixture of oxide or silicates of iron, and points with certainty to the presence of ferruginous minerals or their decomposition-products, which often give considerable fertility to the soil.

Other varieties of sand are not uncommon. In New Jersey occur extensive deposits of so-called *green sand*, containing grains of a mineral, *glauconite*, to be hereafter noticed as a fertilizer. *Lime sand*, consisting of grains of carbonate of lime, is of frequent occurrence on the shores of coral islands or reefs. The term sandy-soil is obviously very indefinite, including nearly the extremes of fertility and barrenness, and covering a wide range of variety as regards composition. It is therefore qualified by various epithets, as coarse, fine, etc. Coarse, sandy soils are usually unprofitable, while fine, sandy soils are often valuable.

Clayey Soils are those in which clay or impalpable matters predominate. They are commonly characterized by extreme fineness of texture, and by great retentive power for water; this liquid finding passage through their pores with extreme slowness. When dried, they become cracked and rifted in every direction from the shrinking that takes place in this process.

It should be distinctly understood that a soil may be clayey without being clay, i. e., it may have the external, physical properties of adhesiveness and impermeability to water which usually characterize clay, without containing those compounds (kaolinite and the like) which constitute clay in the true chemical sense.

On the other hand it were possible to have a soil consisting chemically of clay, which should have the physical properties of sand; for kaolinite has been found in crystals $\frac{1}{3000}$ of an inch in breadth, and destitute of all cohesiveness or plasticity. Kaolinite in such a coarse form is, however, extremely rare, and not likely to exist in the soil.

Loamy Soils are those intermediate in character between sandy and clayey, and consist of mixtures of sand with clay, or of coarse with impalpable matters. They are free from the excessive tenacity of clay, as well as from the too great porosity of sand.

The gradations between sandy and clayey soils are roughly expressed by such terms and distinctions as the following:

KINDS OF SOILS.

	Clay or impalpable matters.	Sand.
Heavy clay contains	75—90%	10—25%
Clay loam "	60—75	25—40
Loam "	40—60	40—60
Sandy loam "	25—40	60—75
Light sandy loam contains	10—25	75—90
Sand "	0—10	90—100

The percentage composition above given applies to the *dry soil*, and must be received with great allowance, since the transition from fine sand to impalpable matter not physically distinguishable from clay, is an imperceptible one, and therefore not well admitting of nice discrimination.

It is furthermore not to be doubted that the difference between a clayey soil and a loamy soil depends more on the form and intimacy of admixture of the ingredients, than upon their relative proportions, so that a loam may exist which contains less sand than some clayey soils.

Calcareous or Lime Soils are those in which *carbonate of lime* is a predominating or characteristic ingredient. They are recognizable by effervescing vigorously when drenched with an acid. Strong vinegar answers for testing them. They are not uncommon in Europe, but in this country are comparatively rare. In the Northern and Middle States, calcareous soils scarcely occur to an extent worthy of mention.

While lime soils exist containing 75%, and more of carbonate of lime, this ingredient is in general subordinate to sand and clay, and we have therefore *calcareous sands*, *calcareous clays*, or *calcareous loams*.

Marls are mixtures of clay or clayey matters, with finely divided carbonate of lime, in something like equal proportions.*

Peat or Swamp Muck is humus resulting from decayed

---

* In New Jersey, *green sand marl*, or marl simply, is the name applied to the green sand employed as a fertilizer. *Shell marl* is a name designating nearly pure carbonate of lime found in swamps.

vegetable matter in bogs and marshes. A soil is peaty or mucky when containing vegetable remains that have suffered partial decay under water.

Vegetable Mold is a soil containing much organic matter that has decayed without submergence in water, either resulting from the leaves, etc., of forest trees, from the roots of grasses, or from the frequent application of large doses of strawy manures.

Ochery or Ferruginous Soils are those containing much oxide or silicates of iron; they have a yellow, red, or brown color.

Other divisions are current among practical men, as, for example, surface and subsoil, active and inert soil, tilth, and hard pan. These terms mostly explain themselves. When, at the depth of four inches to one foot or more, the soil assumes a different color and texture, these distinctions have meaning.

The surface soil, active soil, or *tilth,* is the portion that is wrought by the instruments of tillage—that which is moistened by the rains, warmed by the sun, permeated by the atmosphere, in which the plant extends its roots, gathers its soil-food, and which, by the decay of the subterranean organs of vegetation, acquires a content of humus.

Subsoil.—Where the soil originally had the same characters to a great depth, it often becomes modified down to a certain point, by the agencies just enumerated, in such a manner that the eye at once makes the distinction into surface soil and subsoil. In many cases, however, such distinctions are entirely arbitrary, the earth changing its appearance gradually or even remaining uniform to a considerable depth. Again, the surface soil may have a greater downward extent than the active soil, or the tilth may extend into the subsoil.

Hard pan is the appropriate name of a dense, almost impenetrable, crust or stratum of ochery clay or compacted gravel, often underlying a fairly fruitful soil. It is the soil reverting to rock. The particles once disjointed are being cemented together again by the solutions of lime, iron, or alkali-silicates and humates that descend from the surface soil. Such a stratum often separates the surface soil from a deep gravel bed, and peat swamps thus exist in basins formed on the most porous soils by a thin layer of *moor-bed-pan*.

With these general notions regarding the origin and characters of soils, we may proceed to a somewhat extended notice of the properties of the soil as influencing fertility. These divide themselves into *physical characters*—those which externally affect the growth of the plant; and *chemical characters*—those which provide it with food.

CHAPTER IV.

PHYSICAL CHARACTERS OF THE SOIL.

The physical characters of the soil are those which concern the form and arrangement of its visible or palpable particles, and likewise include the relations of these particles to each other, and to air and water, as well as to the forces of heat and gravitation. Of these physical characters we have to notice:

1. The Weight of Soils.
2. State of Division.
3. Absorbent Power for Vapor of Water, or Hygroscopic Capacity.
4. Property of Condensing Gases.
5. Power of fixing Solid Matters from their Solutions.
6. Permeability to Liquid Water. Capillary Power.
7. Changes of Bulk by Drying, etc.
8. Adhesiveness.
9. Relations to Heat.

In treating of the physical characters of the soil, the writer employs an essay on this subject, contributed by him to Vol. XVI of the Transactions of the N. Y. State Agricultural Society, and reproduced in altered form in a Lecture given at the Smithsonian Institution, Dec., 1859.

§ 1.

THE WEIGHT OF SOILS.

The Absolute Weight of Soils varies directly with their porosity, and is greater the more gravel and sand they contain. In the following Table is given the weight per cubic foot of various soils according to Schübler, and likewise (in round numbers) the weight per acre taken to the depth of one foot (=43,560 cubic feet).

WEIGHT OF SOILS

	per cubic foot	per acre to depth of one foot.
Dry silicious or calcareous sand	about 110 lbs.	4,792,000
Half sand and half clay	" 96 "	4,182,000
Common arable land *	" 80 to 90 "	3,485,000 to 3,920,000
Heavy clay	" 75 "	3,267,000
Garden mold, rich in vegetable matter	" 70 "	3,049,000
Peat soil	" 30 to 50 "	1,307,000 to 2,178,000

From the above figures we see that sandy soils, which are usually termed "light," because they are worked most easily by the plow, are, in fact, the heaviest of all; while clayey land, which is called "heavy," weighs less, bulk for bulk, than any other soils, save those in which vegetable matter predominates. The resistance offered by soils in tillage is more the result of adhesiveness than of gravity. Sandy soils, though they contain in general a less percentage of nutritive matters than clays, may really offer as good

---

* The author is indebted to Prof. Seely, of Middlebury, Vt., for a sample of one-fourth of a cubic foot of Wheat Soil from South Onondaga, New York. The cubic foot of this soil, when dry, weighs 86½ lbs. The acre to depth of one foot weighs 3,768,000 lbs. This soil contains a large proportion of slaty gravel. A rich garden soil of silicious sand that had been heavily dunged, time out of mind. Boussingault found to weigh 81 lbs. av. per cubic foot (1.3 kilos per liter). This would be per acre, one foot deep, 3,528,000 lbs.

nourishment to crops as the latter, since they present one-half more absolute weight in a given space.

Peat soils are light in both senses in which this word is used by agriculturists.

The Specific Gravity of Soils is the weight of a given bulk compared with the same bulk of water. A cubic foot of water weighs 62½ lbs., but comparison of this number with the numbers stated in the last table expressing the weights of a cubic foot of various soils does not give us the true specific gravity of the latter, for the reason that these weights are those of the matters of the soil contained in a cubic foot, but not of a cubic foot of these matters themselves exclusive of the air, occupying their innumerable interspaces. When we exclude the air and take account only of the soil, we find that all soils, except those containing very much humus, have nearly the same density. Schöne has recently determined with care the specific gravity of 14 soils, and the figures range from 2.53 to 2.71. The former density is that of a soil rich in humus, from Orenberg, Russia; the latter of a lime soil from Jena. The density of sandy and clayey soils free from humus is 2.65 to 2.69. (*Bulletin de la Soc. Imp. des Naturalistes de Moscou*, 1867, p. 404.) This agrees with the density of those minerals which constitute the bulk of most soils, as seen from the following statement of their specific gravity, which is, for quartz, 2.65; feldspar, 2.62; mica, 2.75–3.10; kaolinite, 2.60. Calcite has a sp. gr. of 2.72; hence the greater density of calcareous soils.

§ 2.

STATE OF DIVISION OF THE SOIL AND ITS INFLUENCE ON FERTILITY.

On the surface of a block of granite only a few lichens and mosses can exist; crush the block to a coarse powder and a more abundant vegetation can be supported on it;

if it is reduced to a very fine dust and duly watered, even the cereal grains will grow and perfect fruit on it.

Magnus (*Jour. für prakt. Chem.*, L, 70) caused barley to germinate in pure feldspar, which was in one experiment coarsely, in another finely, pulverized. In the coarse feldspar the plants grew to a height of 15 inches, formed ears, and one of them ripened two perfectly formed seeds. In the fine feldspar the plants were very decidedly stronger. One of them attained a height of 20 inches, and produced four seeds.

It is true, as a general rule, that all fertile soils contain a large proportion of fine or impalpable matter. The soil of the "Ree Ree Bottom," on the Scioto River, Ohio, remarkable for its extraordinary fertility, which has remained nearly undiminished for 60 years, though yielding heavy crops of wheat and maize without interruption, is characterized by the fineness of its particles. (D. A. Wells, *Am. Jour. Sci.*, XIV, 11.) In what way the extreme division of the particles of the soil is connected with its fertility is not difficult to understand. The food of the plant as existing in the soil must pass into solution either in the moisture of the soil, or in the acid juices of the roots of plants. In either case the rapidity of its solution is in direct ratio to the extent of surface which it exposes. The finer the particles, the more abundantly will the plant be supplied with its necessary nourishment. In the Scioto valley soils, the water which surrounds the roots of the crops and the root-fibrils themselves come in contact with such an extent of surface that they are able to dissolve the soil-ingredients in as large quantity and as rapidly as the crop requires. In coarse-grained soils this is not so likely to be the case. Soluble matters (manures) must be applied to them by the farmer, or his crops refuse to yield handsomely.

It is furthermore obvious, that, other things being equal, the finer the, articles of the soil the more space the growing roots have in which to expand themselves, and the more abundantly are they able to present their absorbent surfaces to the supplies which the soil contains. The fineness of the particles may, however, be excessive. They may fit each other so closely as to interfere with the growth of the roots, or at least with the sprouting of the seed. The soil may be too compact.

It will presently appear that other very important properties of the soil are more or less related to its state of mechanical division.

§ 3.

ABSORPTION OF VAPOR OF WATER BY THE SOIL.

The soil has a power of withdrawing vapor of water from the air and condensing the same in its pores. It is, in other words, hygroscopic.

This property of a soil is of the utmost agricultural importance, because, 1st, it is connected with the permanent moisture which is necessary to vegetable existence; and, 2d, since the absorption of water-vapor to some degree determines the absorption of other vapors and gases.

In the following table we have the results of a series of experiments carried out by Schübler, for the purpose of determining the absorptive power of different kinds of earths and soils for vapor of water.

The column of figures gives in thousandths the quantity of hygroscopic moisture absorbed in twenty-four hours by the previously dried soil from air confined over water, and hence nearly saturated with vapor.

Quartz sand, coarse	0
Gypsum	1
Lime sand	2
Plough land	23
Clay soil, (60 per cent clay)	28
Slaty marl	33
Loam	35

Fine carbonate of lime	35
Heavy clay soil, (80 per cent clay)	41
Garden mold, (7 per cent humus)	52
Pure clay	49
Carbonate of magnesia (fine powder)	82
Humus	120

Davy found that one thousand parts of the soils named below, after having been dried at 212°, absorbed during one hour of exposure to the air, quantities of moisture as follows:

Sterile soil of Bagshot heath	3
Coarse sand	8
Fine sand	11
Soil from Mersey, Essex	13
Very fertile alluvium, Somersetshire	16
Extremely fertile soil of Ormiston, East Lothian	18

An obvious practical result follows from the facts expressed in the above tables, viz.: that sandy soils which have little attractive force for watery vapor, and are therefore dry and arid, may be meliorated in this respect by admixture with clay, or better with humus, as is done by dressing with vegetable composts and by green manuring. The first table gives us proof that gypsum does not exert any beneficial action in consequence of directly attracting moisture. Humus, or decaying vegetable matter, it will be seen, surpasses every other ingredient of the soil in absorbing vapor of water. This is doubtless in some degree connected with its extraordinary porosity or amount of surface. How the extent of surface alone may act is made evident by comparing the absorbent power of carbonate of lime in the two states of sand and of an impalpable powder. The latter, it is seen, absorbed twelve times as much vapor of water as the former. Carbonate of magnesia stands next to humus, and it is worthy of note that it is a very light and fine powder.

Finally, it is a matter of observation that "silica and lime in the form of *coarse sand* make the soil in which they predominate so dry and hot that vegetation perishes from want of moisture; when, however, they occur as *fine dust*, they form too wet a soil, in which plants suffer from the opposite cause."—(*Hamm's Landwirthschaft.*)

Every body has a definite power of condensing moisture upon its surface or in its pores. Even glass, though presenting to the eye a perfectly clean and dry surface, is coated with a film of moisture. If a piece of glass be weighed on a very delicate balance, and then be wiped with a clean cloth, it will be found to weigh perceptibly less than before. Exposed to the air for an hour or more, it recovers the weight which it had lost by wiping; this loss was water. (Stas. Magnus.) The surface of the glass is thus proved to exert towards vapor of water an adhesive attraction.

Certain compounds familiar to the chemist attract water with great avidity and to a large extent. Oil of vitriol, phosphoric acid, and chloride of calcium, gain weight rapidly when exposed to moist air, or when placed contiguous to other substances which are impregnated with moisture.

For this reason these compounds are employed for purposes of drying. Air, for example, is perfectly freed from vapor of water by slowly traversing a tube containing lumps of dried chloride of calcium, or phosphoric acid, or by bubbling repeatedly through oil of vitriol contained in a suitable apparatus.

Solid substances, which, like chloride of calcium, carbonate of potash, etc., gather water from the air to such an extent as to become liquid, are said to *deliquesce* or to be *deliquescent*. Certain compounds, such as urea, the characteristic ingredient of human urine, deliquesce in moist air and dry away again in a warm atmosphere.

Allusion has been made in "How Crops Grow," p. 55, to the hygroscopic water of vegetation, which furnishes another striking illustration of the condensation of water in porous bodies.

The absorption of vapor of water by solid bodies is not

only dependent on the nature of the substance and its amount of surface, but is likewise influenced by external conditions.

The *rapidity* of absorption depends upon the amount of vapor present or accessible, and is greatest in moist air.

The *amount* of absorption is determined solely by temperature, as Knop has recently shown, and is unaffected by the relative abundance of vapor: i. e., at a given temperature a dry soil will absorb the same amount of moisture from the air, no matter whether the latter be slightly or heavily impregnated with vapor, but will do this the more speedily the more moist the surrounding atmosphere happens to be.

In virtue of this hygroscopic character, the soil which becomes dry superficially during a hot day gathers water from the atmosphere in the cooler night time, even when no rain or dew is deposited upon it.

In illustration of the influence of temperature on the quantity of water absorbed, as vapor, by the soil, we give Knop's observations on a sandy soil from Moeckern, Saxony:

1,000 parts of this soil absorbed
    At 55° F. 13 parts of hygroscopic water.
    " 66° " 11.9 " " " "
    " 77° " 10.2 " " " "
    " 88° " 8.7 " " " "

Knop calculates on the basis of his numerous observations that hair and wool, which are more hygroscopic than most vegetable and mineral substances, if allowed to absorb what moisture they are capable of taking up, contain the following quantities of water, *per cent*, at the temperatures named:

    At 87° Fah., 7.7 per cent.
    " 55° " 15.5 " "
    " 32° " 19.3 " "

Silk is sold in Europe by weight with suitable allowance for hygroscopic moisture, its variable content of which is carefully determined by experiment in each important transaction. It is plain that the circumstances of sale may affect the weight of wool to 10 or more per cent.

§ 4.

CONDENSATION OF GASES BY THE SOIL.

Adhesion.—In the fact that soils and porous bodies generally have a physical absorbing power for the vapor of water, we have an illustration of a principle of very wide application, viz., *The surfaces of liquid and solid matter attract the particles of other kinds of matter.*

This force of *adhesion,* as it is termed, when it acts upon *gaseous* bodies, overcomes to a greater or less degree their expansive tendency, and coerces them into a smaller space—condenses them.

Absorbent Power of Charcoal, etc.—Charcoal serves to illustrate this fact, and some of its most curious as well as useful properties depend upon this kind of physical peculiarity. Charcoal is prepared from wood, itself extremely porous,* by expelling the volatile constituents, whereby the porosity is increased to an enormous extent.

When charcoal is kept in a damp cellar, it condenses so much vapor of water in its pores that it becomes difficult to set on fire. It may even take up one-fourth its own weight. When exposed to various gases and volatile matters, it absorbs them in the same manner, though to very unequal extent.

De Saussure was the first to measure the absorbing power of charcoal for gases. In his experiments, boxwood charcoal was heated to redness and plunged under mer-

---

* Mitscherlich has calculated that the cells of a cubic inch of boxwood have no less than 73 square feet of surface.

cury to cool. Then introduced into the various gases named below, it absorbed as many times its bulk of them, as are designated by the subjoined figures:

Ammonia	90	Hydrochloric acid	85
Sulphurous acid	65	Hydrosulphuric acid	55
Protoxide of nitrogen	40	Carbonic acid	35
Oxygen	$9\frac{1}{4}$	Carbonic oxide	$9\frac{1}{2}$
Hydrogen	$1\frac{3}{4}$	Nitrogen	$7\frac{1}{2}$

According to De Saussure, the absorption was complete in 24 hours, except in case of oxygen, where it continued for a long time, though with decreasing energy. The oxygen thus condensed in the charcoal combined with the carbon of the latter, forming carbonic acid.

Stenhouse more lately has experimented in the same direction. From these researches we learn that the power in question is exerted towards different gases with very unequal effect, and that different kinds of charcoal exert very different condensing power.

Stenhouse found that one gramme of dry charcoal absorbed of several gases the number of cubic centimeters given below.

Name of Gas.	Kind of Charcoal.		
	Wood.	Peat.	Animal.
Ammonia	98.5	96.0	43.5
Hydrochloric acid	45.0	60.0	
Hydrosulphuric acid	30.0	28.5	9.0
Sulphurous acid	32.5	27.5	17.5
Carbonic acid	14.0	10.0	5.0
Oxygen	0.8	0.6	0.5

The absorption or solution of gases in water, alcohol, and other liquids, is analogous to this condensation, and those gases which are most condensed by charcoal are in general, though not invariably, those which dissolve most copiously in liquids, (ammonia, hydrochloric acid).

Condensation of Gases by the Soil.—Reichardt and Blumtritt have recently made a minute study of the kind and amount of gases that are condensed in the pores of various solid substances, including soils and some of their ingredients. (*Jour. für prakt. Chem.*, Bd. 98, p. 476.) Their results relate chiefly to these substances as ordinarily occurring exposed to the atmosphere, and therefore more or less moist. The following Table includes the more important data obtained by subjecting the substances to a temperature of 284° F., and measuring and analyzing the gas thus expelled.

Substance:	100 Grams yielded gas in C. C.	10 Vols. yielded vols. gas.	100 Vols. of Gas contained:			
			Nitrogen.	Oxygen.	Carbonic acid.	Carbonic oxide.
Charcoal, air-dry,	164	—	100	0	0	0
" moistened and dried again,	140	59	86	2	9	3
Peat,	162	—	44	5	51	0
Garden soil, moist,	14	20	64	3	24	9
" " air-dry,	38	54	65	2	33	0
Hydrated oxide of iron, air-dry,	375	309	26	4	70	0
Oxide of iron, ignited,	39	52	83	13	4	0
Hydrated alumina, air-dry,	69	82	41	0	59	—
Alumina, dried at 212°,	11	14	83	17	0	—
Clay,	33	—	65	21	14	—
" long exposed to air,	26	39	70	5	25	—
" moistened,	29	35	60	6	34	—
River silt, air-dry,	40	48	68	0	18	14
" " moistened,	24	29	67	0	31	2
" " again dried,	26	30	67	9	16	7
Carbonate of lime (whiting,) 1864,	43	52	100	0	0	—
" " " " 1865,	39	48	74	16	10	—
" " " precipitated, 1864,	65	—	81	19	0	—
" " " " 1865,	51	52	77	15	8	—
Carbonate of magnesia,	729	125	64	7	29	—
Gypsum, pulverized,	17	—	81	19	0	—

From these figures we gather:

1. The gaseous mixture which is contained in the pores of solid substances rarely has the composition of the atmosphere. In but two instances, viz., with *gypsum* and *precipitated carbonate of lime*, were only oxygen and nitrogen absorbed in proportions closely approaching those of the atmosphere.

2. Nitrogen appears to be nearly always absorbed in greater proportion than oxygen, and is greatly condensed in some cases, as by peat, hydrated oxide of iron, and carbonate of magnesia.

3. Oxygen is often nearly or quite wanting, as in charcoal, oxide of iron, alumina, river silt, and whiting.

4. Carbonic acid, though sometimes wanting entirely, is usually abundant in the absorbed gases.

5. In the pores of charcoal and of soils containing decaying organic matters, carbonic acid is often partially replaced by carbonic oxide. The experiments, however, do not furnish proof that this substance is not *formed* under the influence of the high temperature employed (284° F.) in expelling the gases, rather than by incomplete oxidation of organic matters at ordinary temperatures.

6. A substance, when moist, absorbs less gas than when dry. In accordance with this observation, De Saussure noticed that dry charcoal saturated with various gases evolved a good share of them when moistened with water. Ground (and burnt?) coffee, as Babinet has lately stated, evolves so much gas when drenched with water as to burst a bottle in which it is confined.

The extremely variable figures obtained by Blumtritt when operating with the same substance (the figures given in the table are averages of two or three usually discordant results), result from the general fact that the proportion in which a number of gases are present in a mixture, influences the proportion of the individual gases absorbed. Thus while charcoal or soil will absorb a large amount of ammonia from the pure gas, it will take up but traces of this substance from the atmosphere of which ammonia is but an infinitesimal ingredient.

So, too, charcoal or soil saturated with ammonia by exposure to the unmixed gas, loses nearly all of it by standing in the air for some time. This is due to the fact that *gases attract each other*, and the composition of the gas condensed in a porous body varies perpetually with the variations of composition in the surrounding atmosphere.

It is especially the water-gas (vapor of water) which is a fluctuating ingredient of the atmosphere, and one which is absorbed by porous bodies in the largest quantity. This not only displaces other gases from their adhesion to solid surfaces, but by its own attractions modifies these adhesions.

Reichardt and Blumtritt take no account of water-gas, except in the few experiments where the substances were purposely moistened. In all their trials, however, moisture was present, and had its quantity been estimated, doubtless its influence on the extent and kind of absorption would have been strikingly evident throughout.

Ammonia and carbonate of ammonia in the gaseous form are absorbed from the air by the dry soil, to a less degree than by a soil that is moist, as will be noticed fully hereafter.

Chemical Action induced by Adhesion.—This physical property often leads to remarkable chemical effects; in other words, adhesion exalts or brings into play the force of affinity. When charcoal absorbs those emanations from putrefying animal matters which we scarcely know, save by their intolerable odor and poisonous influence, it causes at the same time their rapid and complete oxidation; and hence a piece of tainted meat is sweetened by covering it with a thin layer of powdered charcoal. As Stenhouse has shown, the carcass of a small animal may be kept in a living-room during the hottest weather without giving off any putrid odor, provided it be surrounded on all sides by a layer of powdered charcoal an inch or more thick. Thus circumstanced, it simply smells of *ammonia*, and its destructible parts are resolved directly into water, carbonic acid, free nitrogen, and ammonia, precisely as if they were burned in a furnace, and without the appearance of any of the effluvium that ordinarily arises from decaying flesh.

The metal platinum exhibits a remarkable condensing power, which is manifest even with the polished surface of foil or wire; but is most striking when the metal is brought to the condition of sponge, a form it assumes when certain of its compounds (e. g. ammonia-chloride of platinum) are decomposed by heat, or to the more finely divided state of platinum black. The latter is capable of condensing from 100 to 250 times its volume of oxygen, according to its mode of preparation (its porosity?); and for this reason it possesses intense oxidizing power, so that, for example, when it is brought into a mixture of oxygen and hydrogen, it causes them to unite explosively. A jet of hydrogen gas, allowed to play on platinum sponge, is almost instantly ignited—a fact taken advantage of in Döbereiner's hydrogen lamp.

The oxidizing powers of platinum are much more vigorous than those of charcoal. Stenhouse has proposed the use of platinized charcoal (charcoal ignited after moistening with solution of chloride of platinum) as an escharotic and disinfectant for foul ulcers, and has shown that the foul air of sewers and vaults is rendered innocuous when filtered or breathed through a layer of this material.*

Chemical Action a Result of the Porosity of the Soil.—From these significant facts it has been inferred that the soil by virtue of the extreme porosity of some of its ingredients is the theater of chemical changes of the utmost importance, which could not transpire to any sensible extent but for this high division of its particles and the vast surface they present.

The soil absorbs putrid and other disagreeable effluvia, and undoubtedly oxidizes them like charcoal, though, perhaps, with less energy than the last named substance, as would be anticipated from its inferior porosity. Garments which have been rendered disgusting by the fetid secretions of the skunk, may be "sweetened," i. e. deprived of

---

* Platinum does not condense hydrogen gas; but the metal *Palladium*, which occurs associated with platinum, has a most astonishing absorptive power for hydrogen, being able to take up or "occlude" 900 times its volume of the gas. (Graham, *Proceedings Roy. Soc.*, 1868, p. 422.)

odor, by burying them for a few days in the earth. The Indians of this country are said to sweeten the carcass of the skunk by the same process, when needful, to fit it for their food. Dogs and foxes bury bones and meat in the ground, and afterward exhume them in a state of comparative freedom from offensive odor.

When human excrements are covered with fine dry earth, as in the "Earth Closet" system, all odor is at once suppressed and never reappears. At the most, besides an "earthy" smell, an odor of ammonia appears, resulting from decomposition, which appears to proceed at once to its ultimate results without admitting of the formation of any intermediate offensive compounds.

Dr. Angus Smith, having frequently observed the presence of nitrates in the water of shallow town wells, suspected that the nitric acid was derived from animal matters, and to test this view, made experiments on the action of filters of sand, and other porous bodies, upon solutions of different animal and vegetable matters. He found that in such circumstances oxidation took place most rapidly—the nitrogen of organic matters being converted into nitric acid, the carbon and hydrogen combining with oxygen at the same time. Thus a solution of yeast, which contained no nitric acid, after being passed through a filter of sand, gave abundant evidence of salts of this acid. Colored solutions were in this way more or less decolorized. Water, rendered brown by peaty matter, was found to be purified by filtration through sand.*

§ 5.
POWER OF SOILS TO REMOVE DISSOLVED SOLIDS FROM THEIR SOLUTIONS.

Action of Sand upon Saline Solutions.—It has long been known that simple *sand* is capable of partially re-

---

* This account of Dr. Smith's experiments is quoted from Prof. Way's paper "On the Power of Soils to Absorb Manure." (*Jour. Roy. Ag. Soc. of England*, XI, p. 317.)

moving saline matters from their solutions in water. Lord Bacon, in his "Sylva Sylvarum," speaks of a method of obtaining fresh water, which was practised on the coast of Barbary. "Digge a hole on the sea-shore somewhat above high-water mark and as deep as low-water mark, which, when the tide cometh, will be filled with water fresh and potable." He also remarks "to have read that trial hath been made of salt-water passed through earth through ten vessels, one within another, and yet it hath not lost its saltness as to become potable;" but when "draymed through twenty vessels, hath become fresh."

Dr. Stephen Hales, in a paper read before the Royal Society in 1739, on "Some attempts to make sea-water wholesome," mentions on the authority of Mr. Boyle Godfrey that "sea-water, being filtered through stone cisterns, the first pint that runs through will be pure water having no taste of the salt, but the next pint will be salt as usual."

Berzelius found upon filtering solutions of common salt through sand, that the portions which first passed were quite free from saline impregnation. Matteucci extended this observation to other salts, and found that the solutions when filtered through sand were diminished in density, showing a detention by the sand of certain quantities of the salt operated upon.*

Action of Humus on Saline Solutions.—Heiden (*Hoff-mann's Jahresbericht*, 1866, p. 29) found that peat and various preparations of the humic acids, when brought into solutions of chloride of potassium and chloride of ammonium, remove a portion of these salts from the liquid, leaving the solutions perceptibly weaker. The removed salts were for the most part readily dissolved by a small quantity of water. W. Schumacher (*Hoff. Jahres.*, 1867, p. 18) observed that humus, artificially prepared by the

---

* These statements of Bacon, Hales, Berzelius, and Matteucci, are derived from Prof. Way's paper "On the Power of Soils, etc." (*Jour. Roy. Ag. Soc. of Eng.*, XI, 316.)

action of oil of vitriol on sugar, when placed in ten times its quantity of solutions of various salts (containing about ½ per cent of solid matter) absorbed of sulphates of soda and ammonia, and chlorides of calcium and ammonium, about 2 per cent; of sulphate of potash 4 per cent; and of phosphate of soda 10 per cent. Schumacher also noticed that sulphate of potash is able to expel sulphate of ammonia from humic acid which has been saturated with the latter salt, but that the latter cannot displace the former. In Schumacher's experiments, pure water freely dissolved the salts absorbed by the humic acid.

Explanation.—Let us consider what occurs in the act of solution and in this separation of soluble matters from a liquid. The difference between the solid and the liquid state, so far as we can define it, lies in the unequal cohesion of the particles. Cohesion prevails in solids, and opposes freedom of motion among the particles. In liquids, cohesion is not altogether overcome but is greatly weakened, and the particles move easily upon each other. When a lump of salt is put into water, the cohesion that otherwise maintains its particles in the solid state is overcome by the attraction of adhesion, which is mutually exerted between them and the particles of water, and the salt dissolves. If now into the solution of salt any insoluble solid be placed which the liquid can wet (adhere to) its particles will exert adhesive attraction for the particles of salt, and the tendency of the latter will be to concentrate somewhat upon the surface of the solid.

If the solid, thus introduced into a solution, be exceedingly porous, or otherwise present a great amount of surface, as in case of sand or humus, this tendency is proportionately heightened, and a separation of the dissolved substance may become plainly evident on proper examination. When, on the other hand, the solid surface is relatively small, no weakening of the solution may be perceptible by ordinary means. Doubtless the glass of a bottle containing brine concentrates the latter where the two are in contact, though the effect may be difficult to demonstrate.

Defecating Action of Charcoal on Solutions.—Charcoal manifests a strong surface attraction for various solid substances, and exhibits this power by overcoming the adhesion they have to the particles of water when dissolved in that fluid. If ink, solution of indigo, red wine, or bitter ale, be agitated some time with charcoal, the color, and in the case of ale, the bitter principle, will be taken up by the charcoal, leaving the liquid colorless and comparatively tasteless. Water, which is impure from putrefying organic matters, is sweetened, and brown sugars are whitened by the use of charcoal or bone-black. In case of bone-black, the finely divided bone-earth (phosphate of lime) assists the action of the charcoal.

Fixing of Dye-Stuffs.—The familiar process of dyeing depends upon the adhesion of coloring matters to the fiber of textile fabrics. Wool steeped in solution of indigo attaches the pigment permanently to its fibers. Silk in the same way fastens the particles of rosaniline, which constitutes the magenta dye. Many colors, e. g. madder and logwood, which will not adhere themselves directly to cloth, are made to dye by the use of mordants—substances like alumina, oxide of tin, etc.—which have adhesion both to the fabric and the pigment.

Absorptive Power of Clay.—These effects of charcoal and of the fibers of cotton, etc., are in great part identical with those previously noticed in case of sand and humus. Their action is, however, more intense, and the effects are more decided. Charcoal, for example, that has absorbed a pigment or a bitter principle from a liquid, will usually yield it up again to the same or a stronger solvent. In some instances, however, as in dyeing with simple colors, matters are fixed in a state of great permanence by the absorbent; and in others, as where mordants are used, chemical combinations supervene, which possess extraordinary stability.

Many facts are known which show that soils, or certain of their ingredients, have a fixing power like that of charcoal and textile fibers. It is a matter of common experience that a few feet or yards of soil intervening between a cess-pool or dung-pit, and a well, preserves the latter against contamination for a longer or shorter period.

J. P. Bronner, of Baden, in a treatise on "Grape Culture in South Germany," published in 1836, first mentions that dung liquor is deodorized, decolorized, and rendered nearly tasteless by filtration through garden earth. Mr. Huxtable, of England, made the same observation in 1848, and Prof. Way and others have published extended investigations on this extremely important subject.

Prof. Way informs us that he filled a long tube to the depth of 18 inches with Mr. Huxtable's light soil, mixed with its own bulk of white sand. "Upon this filter-bed a quantity of highly offensive stinking tank water was poured. The liquid did not pass for several hours, but ultimately more than 1 ounce of it passed *quite clear*, free from smell or *taste*, except a peculiar earthy smell and taste derived from the soil." Similar results were obtained by acting upon putrid human urine, upon the stinking water in which flax had been steeped, and upon the water of a London sewer.

Prof. Way found that these effects were not strikingly manifested by pure sand, but appeared when clay was used. He found that solutions of coloring matters, such as logwood, sandal-wood, cochineal, litmus, etc., when filtered through or shaken up with a portion of clay, are entirely deprived of color. (*Jour. Roy. Ag. Soc. of Eng.*, XI, p. 364.)

These effects of clay or clayey matters, like the fixing power of cotton and woolen stuffs upon pigments, must be regarded for the most part as purely physical. There are other results of the action of the soil on saline solutions, which, though perhaps influenced by simple physical action, are preponderatingly chemical in their aspect. These effects, which manifest themselves by chemical decompositions and substitutions, will be fully discussed in a subsequent chapter, p. 333.

§ 6.
PERMEABILITY OF SOILS TO LIQUID WATER. IMBIBITION. CAPILLARY POWER.

The fertility of the soil is greatly influenced by its deportment toward water in the liquid state.

A soil is *permeable* to water when it allows that liquid to soak into or run through it. To be permeable is of course to be porous. On the size of the pores depends its degree of permeability. Coarse sands, and soils which have *few* but *large* pores or interspaces, allow water to run through them readily—water *percolates* them. When, instead of running through, the water is largely absorbed and held by the soil, the latter is said to possess great *capillary power;* such a soil has *many* and *minute* pores. The cause of capillarity is the same surface attraction which has been already under notice.

When a narrow vial is partly filled with water, it will be seen that the liquid adheres to its sides, and if it be not more than one-half inch in diameter, the surface of the liquid will be curved or concave. In a very narrow tube the liquid will rise to a considerable height. In these cases the surface attraction of the glass for the water neutralizes or overcomes the weight of (earth's attraction for) the latter.

The pores of a sponge raise and hold water in them, in the same way that these narrow (capillary*) tubes sup-

---

* From *capillus*, the Latin word for hair, because as fine as hair; (but a hair is no tube, as is often supposed.)

port it. When a body has pores so fine (surfaces so near each other) that their surface attraction is greater than the gravitating tendency of water, then the body will imbibe and hold water—will exhibit capillarity; a lump of salt or sugar, a lamp-wick, are familiar examples. When the pores of a body are so large (the surfaces so distant) that they cannot fill themselves or keep themselves full, the body allows the water to run through or to percolate.

Sand is most easily permeable to water, and to a higher degree the coarser its particles. Clay, on the other hand, is the least penetrable, and the less so the purer and more plastic it is.

When a soil is too coarsely porous, it is said to be *leachy* or *hungry*. The rains that fall upon it quickly soak through, and it shortly becomes dry. On such a soil, the manures that may be applied in the spring are to some degree washed down below the reach of vegetation, and in the droughts of summer, plants suffer or perish from want of moisture.

When the texture of a soil is too fine,—its pores too small,—as happens in a heavy clay, the rains penetrate it too slowly; they flow off the surface, if the latter be inclined, or remain as pools for days and even weeks in the hollows.

In a soil of proper texture the rains neither soak off into the under-earth nor stagnate on the surface, but the soil always (except in excessive wet or drought) maintains the moistness which is salutary to most of our cultivated plants.

Movements of Water in the Soil.—If a wick be put into a lamp containing oil, the oil, by capillary action, gradually permeates its whole length, that which is above as well as that below the surface of the liquid. When the lamp is set burning, the oil at the flame is consumed, and as each particle disappears its place is supplied by a new one, until the lamp is empty or the flame extinguished.

Something quite analogous occurs in the soil, by which the plant (corresponding to the flame in our illustration) is fed. The soil is at once lamp and wick, and the *water of the soil* represents the oil. Let evaporation of water from the surface of the soil or of the plant take the place of the combustion of oil from a wick, and the matter stands thus: Let us suppose dew or rain to have saturated the ground with moisture for some depth. On recurrence of a dry atmosphere with sunshine and wind, the surface of the soil rapidly dries; but as each particle of water escapes (by evaporation) into the atmosphere, its place is supplied (by capillarity) from the stores below. The ascending water brings along with it the soluble matters of the soil, and thus the roots of plants are situated in a stream of their appropriate food. The movement proceeds in this way so long as the surface is drier than the deeper soil. When, by rain or otherwise, the surface is saturated, it is like letting a thin stream of oil run upon the apex of the lamp-wick—no more evaporation into the air can occur, and consequently there is no longer any ascent of water; on the contrary, the water, by its own weight, penetrates the soil, and if the underlying ground be not saturated with moisture, as can happen where the subterranean fountains yield a meagre supply, then capillarity will aid gravity in its downward distribution.

It is certain that a portion of the mineral matters, and, perhaps, also some organic bodies which feed the plant, are more or less freely dissolved in the water of the soil. So long as evaporation goes on from the surface, so long there is a constant upward flow of these matters. Those portions which do not enter vegetation accumulate on or near the surface of the ground; when a rain falls, they are washed down again to a certain depth, and thus are kept constantly changing their place with the water, which is the vehicle of their distribution. In regions where rain falls periodically or not at all, this upward flow of the soil-water often causes an accumulation of salts on the surface of the ground. Thus in Bengal many soils which in the wet season produce the most luxuriant crops, during the rainless portion of the year become covered with white crusts of saltpeter. The beds of nitrate of soda that are found in Peru, and the carbonate of soda and other salts which incrust the deserts of Utah, and often fill the air with alkaline dust, have accumulated in the same manner. So in our western caves the earth sheltered from rains is saturated with salts—epsom-salts, Glauber's-salts, and saltpeter, or mixtures of these. Often the rich soil of gardens is slightly incrusted in this manner in our summer weather; but the saline matters are carried into the soil with the next rain.

It is easy to see how, in a good soil, capillarity thus acts in keeping the roots of plants constantly immersed in a stream of water or moisture that is now ascending, now descending, but never at rest, and how the food of the plant is thus made to circulate around the organs fitted for absorbing it.

The same causes that maintain this perpetual supply of water and food to the plant are also efficacious in constantly preparing new supplies of food. As before explained, the materials of the soil are always undergoing decomposition, whereby the silica, lime, phosphoric acid, potash, etc., of the insoluble fragments of rock, become soluble in water and accessible to the plant. Water charged with carbonic acid and oxygen is the chief agent in these chemical changes. The more extensive and rapid the circulation of water in the soil, the more matters will be rendered soluble in a given time, and, other things being equal, the less will the soil be dependent on manures to keep up its fertility.

Capacity of Imbibition. Capillary Power.—No matter how favorable the structure of the soil may be to the circulation of water in it, no continuous upward movement can take place without *evaporation*. The ease and rapidity of evaporation, while mainly depending on the condition of the atmosphere and on the sun's heat, are to a certain degree influenced by the soil itself. We have already seen that the soil possesses a power of absorbing watery vapor from the atmosphere, a power which is related both to the kind of material that forms the soil and to its state of division. This absorptive power opposes evaporation. Again, different soils manifest widely different capacities for *imbibing liquid water*—capacities mainly connected with their porosity. Obviously, too, the quantity of liquid in a given volume of soil affects not only the rapidity, but also the duration of evaporation.

The following tables by Schübler illustrate the peculiarities of different soils in these respects. The first column gives the *percentages* of *liquid water* absorbed by the completely dry soil. In these experiments the soils were thoroughly wet with water, the excess allowed to drip off, and the increase of weight determined. In the second column are given the percentages of water that evaporated during the space of four hours from the saturated soil spread over a given surface:

Quartz sand	25	88.4
Gypsum	27	71.7
Lime sand	29	75.9
Slaty marl	34	68.0
Clay soil, (sixty per cent clay,)	40	52.0
Loam	51	45.7
Plough land	52	32.0
Heavy clay, (eighty per cent clay,)	61	34.9
Pure gray clay	70	31.9
Fine carbonate of lime	85	28.0
Garden mould	89	24.3
Humus	181	25.5
Fine carbonate of magnesia	256	10.8

It is obvious that these two columns express nearly the same thing in different ways. The amount of water retained increases from quartz sand to magnesia. The rapidity of drying in the air diminishes in the same direction.

Some observations of Zenger (*Wilda's Centralblatt*, 1858, 1, 430) indicate the influence of the state of division of a soil on its power of imbibing water. In the subjoined table are given in the first column the per cent of water imbibed by various soils which had been brought to nearly the same degree of moderate fineness by sifting off both the coarse and the fine matter; and the second column gives the amounts imbibed by the same soils, reduced to a high state of division by pulverization.

	Coarse.	Fine.
Quartz sand,	26.0	53.5
Marl (used as fertilizer,)	30.2	54.5
Marl, underlying peat,	39.0	48.5
Brick clay,	66.2	57.5
Moor soil,	104.5	101.0
Alm (lime-sinter,)	108.3	70.4
Alm soil,	178.2	102.5
Peat dust,	377.0	268.5

The effects of pulverization on soils whose particles are compact is to increase the surface, and increase to a corresponding degree the imbibing power. On soils consisting of porous particles, like lime-sinter and peat, pulverization destroys the porosity to some extent and diminishes the amount of absorption. The first class of soils are probably increased in bulk, the latter reduced, by grinding.

Wilhelm, (*Wilda's Centralblatt*, 1866, 1, 118), in a series of experiments on various soils, confirms the above results of Zenger. He found, e. g., that a garden mould imbibed 114 per cent, but when pulverized absorbed but 62 per cent.

To illustrate the different properties of various soils for which the farmer has but one name, the fact may be adduced that while Schübler, Zenger, and Wilhelm found the imbibing power of "clay" to range between 40 and 70 per cent, Stoeckhardt examined a "clay" from Saxony that held 150 per cent of water. So the humus of Schübler imbibed 181 per cent; the peat of Zenger, 377 per cent; while Wilhelm examined a very porous peat that took up 519 per cent. These differences are dependent mainly on the mechanical texture or porosity of the material.

The want of capillary retentive power for water in the case of coarse sand is undeniably one of the chief reasons of its unfruitfulness. The best soils possess a medium retentive power. In them, therefore, are best united the conditions for the regular distribution of the soil-water under all circumstances. In them this process is not hindered too much either by wet or dry weather. The retaining power of humus is seen to be more than double that of clay. This result might appear at first sight to be in contradiction to ordinary observations, for we are accustomed to see water standing on the surface of clay but not on humus. It must be borne in mind that clay, from its imperviousness, holds water like a vessel, the water remaining apparent; but humus retains it invisibly, its action being nearly like that of a sponge.

One chief cause of the value of a layer of humus on the surface of the soil doubtless consists in this great retaining power for water, and the success that has attended the practice of green manuring, as a means of renovating almost worthless shifting sands, is in a great degree to be attributed to this cause. The advantages of mulching are explained in the same way.

Soils which are over-rich in humus, especially those of reclaimed peat-bogs, have some detrimental peculiarities deserving notice. Stoeckhardt (*Wilda's Centralblatt*, 1858, 2, 22) examined the soil of a cultivated moor in Saxony, which, when moist, had an imbibing power of 60–69°$|_0$. After being thoroughly dried, however, it lost its adhesiveness, and the imbibing power fell to 26–30°$|_0$.

It is observed in accordance with these data that such soils retain water late in spring; and when they become very dry in summer they are slow to take up water again, so that rain-water stands on the surface for a considerable time without penetrating, and when, after some days, it is soaked up, it remains injuriously long. Light rains *after drought* do little immediate good to such soils, while heavy rains always render them too wet and cold, unless they are suitably ameliorated. The same is true to a less degree of heavy, compact clays.

§ 7.

CHANGES OF THE BULK OF THE SOIL BY DRYING AND FROST.

The Shrinking of Soils on Drying is a matter of no little practical importance. This shrinking is of course offset by an increase of bulk when the soil becomes wet. In variable weather we have therefore constant changes of volume occurring.

Soils rich in humus experience these changes to the greatest degree. The surfaces of moors often rise and fall with the wet or dry season, through a space of several inches. In ordinary light soils, containing but little humus, no change of bulk is evident. Otherwise, it is in clay soils that shrinking is most perceptible; since these soils only dry superficially, they do not appear to settle much, but become full of cracks and rifts. Heavy clays may lose one-tenth or more of their volume on drying, and since at the same time they harden about the rootlets which are imbedded in them, it is plain that these indispensable organs of the plant must thereby be ruptured during the protracted dry weather. Sand, on the other hand, does not change its bulk by wetting or drying, and when present to a considerable extent in the soil, its particles, being interposed between those of the clay, prevent the adhesion of the latter, so that, although a sandy loam shrinks not inconsiderably on drying, yet the lines of separation are vastly more numerous and less wide than in purer clays. Such a soil does not "cake," but remains friable and powdery.

Marly soils (containing carbonate of lime) are especially prone to fall to a fine powder during drying, since the carbonate of lime, which, like sand, shrinks very little, is itself in a state of extreme division, and therefore more effectually separates the clayey particles. The unequal shrinking of these two intimately mixed ingredients accomplishes a perfect pulverization of such soils. On the cold, heavy soils of Upper Lusatia, in Germany, the application of lime has been attended with excellent results, and the larger share of the benefit is to be accounted for by the improvement in the texture of those soils which follows liming. The carbonate of lime is considerably soluble in water charged with carbonic acid, as is the water of a soil containing vegetable matter, and this agency of distribution, in connection with the mechanical operations of tillage, must in a short time effect an intimate mixture of the lime with the whole soil. A tenacious clay is thus by a heavy liming made to approach the condition of a friable marl.

Heaving by Frost.—Soils which imbibe much water, especially clay and peat soils, have likewise the disagreeable property of being heaved by frost. The expansion, by freezing, of the liquid water they contain, separates the particles of soil from each other, raises, in fact, the surface for a considerable height, and thus ruptures the roots of grass and especially of fall-sowed grain. The lifting of fence posts is due to the same cause.

§ 8.

ADHESIVENESS OF THE SOIL.

In the language of the farm a soil is said to be heavy or light, not as it weighs more or less, but as it is easy or difficult to work. The *state of dryness* has great influence on this quality. Sand, lime, and humus have very little adhesion when dry, but considerable when wet. Soils in which they predominate are usually easy to work. But clay or impalpable matter has entirely different characters, upon which the tenacity of a soil almost exclusively depends. Dry "clay," when powdered, has hardly more consistence than sand, but when thoroughly moistened its particles adhere together to a soft and plastic, but tenacious mass; and in drying away, at a certain point it becomes very hard, and requires a good deal of force to penetrate it. In this condition it offers great resistance to the instruments used in tillage, and when thrown up by the plow it forms lumps which require repeated harrowings to break them down. Since the adhesiveness of the soil depends so greatly upon the quantity of water contained in it, it follows that thorough draining, combined with deep tillage, whereby sooner or later the stiffest clays become readily permeable to water, must have the best effects in making such soils easy to work.

The English practice of burning clays speedily accomplishes the same purpose. When clay is burned and then crushed, the particles no longer adhere tenaciously together on moistening, and the mass does not acquire again the unctuous plasticity peculiar to unburned clay.

Mixing sand with clay, or incorporating vegetable matter with it, or liming, serves to separate the particles from each other, and thus remedies too great adhesiveness.

The considerable expansion of water in the act of solidifying (one-fifteenth of its volume) has already been noticed as an agency in reducing rocks to powder. In the same way the alternate freezing and thawing of the water which impregnates the soil during the colder part of the year plays an important part in overcoming its adhesion. The effect is apparent in the spring, immediately after "the frost leaves the ground," and is very considerable, fully *one-third* of the resistance of a clay or loam to the plow thus disappearing, according to Schübler's experiments.

Tillage, when carried on with the soil in a wet condition, to some extent neutralizes the effects of frost, especially in tenacious soils.

Fall-plowing of stiff soils has been recommended, in order to expose them to the disintegrating effects of frost.

§ 9.

RELATIONS OF THE SOIL TO HEAT.

The relations of the soil to heat are of the utmost importance in affecting its fertility. The distribution of plants is, in general, determined by differences of mean temperature. In the same climate and locality, however, we find the farmer distinguishing between cold and warm soils.

The Temperature of the Soil varies to a certain depth with that of the air; yet its changes occur more slowly, are confined to a considerably narrower range, and diminish downward in rapidity and amount, until at a certain depth a point is reached where the temperature is invariable.

In summer the temperature of the soil is higher in daytime than that of the air; at night the temperature of the surface rapidly falls, especially when the sky is clear.

In temperate climates, at a depth of three feet, the temperature remains unchanged from day to night; at a depth of 20 feet the annual temperature varies but a degree or two; at 75 feet below the surface, the thermometer remains perfectly stationary. In the vaults of the Paris Observatory, 80 feet deep, the temperature is 50° Fahrenheit. In tropical regions the point of nearly unvarying temperature is reached at a depth of one foot.

The mean annual temperature of the soil is the same as, or in higher latitudes a degree above, that of the air. The nature and position of the soil must considerably influence its temperature

Sources of the Heat of the Soil.—The sources of that heat which is found in the soil are three, viz.: First, the original heat of the earth; second, the chemical process of oxidation or decay going on within it; and third, an external one, the rays of the sun

The earth has within itself a source of heat, which maintains its interior at a high temperature; but which escapes so rapidly from the surface that the soil would be constantly frozen but for the external supply of heat from the sun.

The heat evolved by the decay of organic matters is not inconsiderable in porous soils containing much vegetable remains; but decay cannot proceed rapidly until the external temperature has reached a point favorable to vegetation, and therefore this source of heat probably has no appreciable effect, one way or the other, on the welfare of the plant. The warmth of the soil, so far as it favors vegetable growth, appears then to depend exclusively on the heat of the sun.

The direct rays of the sun are the immediate cause of the warmth of the earth's surface. The temperature of the soil near the surface changes progressively with the seasons; but at a certain depth the loss from the interior and the gain from the sun compensate each other, and, as has been previously mentioned, the temperature remains unchanged throughout the year.

Daily Changes of Temperature.—During the day the sun's heat reaches the earth directly, and is absorbed by the soil and the solid objects on its surface, and also by the air and water. But these different bodies, and also the different kinds of soil, have very different ability to absorb or become warmed by the sun's heat. Air and water are almost incapable of being warmed by heat applied above them. Through the air, heat radiates without being absorbed. Solid bodies which have dull and porous surfaces absorb heat most rapidly and abundantly. The soil and solid bodies become warmed according to their individual capacity, and from them the air receives the heat which warms it. From the moist surface of the soil goes on a rapid evaporation of water, which consumes* a large amount of heat, so that the temperature of the soil is not rapidly but gradually elevated. The ascent of water from the subsoil to supply the place of that evaporated, goes on as before described. When the sun declines, the process diminishes in intensity, and when it sets, the reverse takes place. The heat that had accumulated on

---

* When a piece of ice is placed in a vessel whose temperature is increasing, by means of a lamp, at the rate of one degree of the thermometer every minute, it will be found that the temperature of the ice rises until it attains 32°. When this point is reached, it begins to melt, but does not suddenly become fluid: the melting goes on very gradually. A thermometer placed in the water remains constantly at 32° so long as a fragment of ice is present. The moment the ice disappears, the temperature begins to rise again, at the rate of one degree per minute. The time during which the temperature of the ice and water remains at 32° is 140 minutes. During each of these minutes one degree of heat enters the mixture, but is not indicated by the thermometer—the mercury remains stationary; 140° of heat have thus passed into the ice and become hidden, *latent;* at the same time the solid ice has become liquid water. The difference, then, between ice and water consists in the heat that is latent in the latter. If we now proceed with the above experiment, allowing the heat to increase with the same rapidity, we find that the temperature of the water rises constantly for 180 minutes. The thermometer then indicates a temperature of 212°, (32+180,) and the water boils. Proceeding with the experiment, the water evaporates away, but the thermometer continues stationary so long as any liquid remains. After the lapse of 972 minutes, it is completely evaporated. Water in becoming steam renders, therefore, still another portion, 972°, of heat latent. The heat latent in steam is indispensable to the existence of the latter. If this heat be removed by bringing the steam into a cold space, water is reproduced. If, by means of pressure or cold, steam be condensed, the heat originally latent in it becomes sensible, *free,* and capable of affecting the thermometer. If, also, water be converted into ice, as much heat is evolved and made sensible as was absorbed and made latent. It is seen thus that the processes of liquefaction and vaporization are *cooling* processes; for the heat rendered latent by them must be derived from surrounding objects, and thus these become cooled. On the contrary, solidification, freezing, and vapor-condensation, are *warming* processes, since in them large quantities of heat cease to be latent and are made sensible, thus warming surrounding bodies.

the surface of the earth radiates into the cooler atmosphere and planetary space; the temperature of the surface rapidly diminishes, and the air itself becomes cooler by convection.* As the cooling goes on, the vapor suspended in the atmosphere begins to condense upon cool objects, while its latent heat becoming free hinders the too sudden reduction of temperature. The condensed water collects in drops—it is dew; or in the colder seasons it crystallizes as hoar-frost.

The deposition of liquid water takes place not on the surface of the soil merely, but within it, and to that depth in which the temperature falls during the night, viz., 12 to 18 inches. (Krutzsch observed the temperature of a garden soil at the depth of one foot, to rise 3° F. on a May day, from 9 A. M. to 7 P. M.)

Since the air contained in the interstices of the soil is at a little depth saturated with aqueous vapor, it results that the slightest reduction of temperature must at once occasion a deposition of water, so that the soil is thus supplied with moisture independently of its hygroscopic power.

Conditions that Affect the Temperature of the Soil.—The special nature of the soil is closely connected with the maintenance of a uniform temperature, with the prevention of too great heat by day and cold by night, and with the watering of vegetation by means of dew. It is, however, in many cases only for a little space after seed-time that the soil is greatly concerned in these processes. So soon as it becomes covered with vegetation, the char-

---

* Though liquids and gases are almost perfect non-conductors of heat, yet it can diffuse through them rapidly, if advantage be taken of the fact that by heating they expand and therefore become specifically lighter. If heat be applied to the upper surface of liquids or gases, they remain for a long time nearly unaffected; if it be applied *beneath* them, the lower layers of particles become heated and rise, their place is supplied by others, and so currents upward and downward are established, whereby the heat is rapidly and uniformly distributed. This process of *convection* can rarely have any influence *in* the soil. What we have stated concerning it shows, however, in what way the atmosphere may constantly act in removing heat from the surface of the soil.

acter of the latter determines to a certain degree the nature of the atmospheric changes. In case of many crops, the soil is but partially covered, and its peculiarities are then of direct influence on its temperature.

Relation of Temperature to Color and Texture.—It is usually stated that black or dark-colored soils are sooner warmed by the sun's rays than those of lighter color, and remain constantly of a higher temperature so long as the sun acts on them. An elevation of several degrees in the temperature of a light-colored soil may be caused by strewing its surface with peat, charcoal powder, or vegetable mould. To this influence may be partly ascribed the following facts. Lampadius was able to ripen melons, even in the coolest summers, in Freiberg, Saxony, by strewing a coating of coal dust an inch deep over the surface of the soil. In Belgium and on the Rhine, it is found that the grape matures best, when the soil is covered with fragments of black clay slate.

According to Creuzé-Latouche, the vineyards along the river Loire grow either upon a light-colored calcareous soil, or upon a dark red earth. These two kinds of soil often alternate with each other within a little distance, and the character of the wine produced on them is remarkably connected with the color of the earth. On the light-colored soils only a weak, white wine can be raised to advantage, while on contiguous dark soils a strong claret of fine quality is made. (Gasparin, *Cours d' Agriculture*, 1, 103.)

Girardin found in a series of experiments on the cultivation of potatoes, that the time of their ripening varied eight to fourteen days, according to the color of the soil. He found on August 25th, in a very dark humus soil, twenty-six varieties ripe; in sandy soil, twenty; in clay, nineteen; and in white lime soil, only sixteen. It is not difficult, however, to indicate other causes that will account in part for the results of Girardin.

Schübler made observations on the temperatures attained by various dry soils exposed to the sun's rays, according as their surfaces were blackened by a thin sprinkling of lamp-black or whitened by magnesia. His results are given in columns 1 and 2 of the following table (*vide* p. 196,) from which it is seen that the dark surface was warmed 13° to 14° more than the white. We likewise notice that the character of the very *surface* determines the degree of warmth, for, under a sprinkling of lamp-black or magnesia, all the soils experimented with became as good as identical in their absorbing power for the sun's heat.

The observations of Malaguti and Durocher prove that the peculiar temperature of the soil is not always so closely related to color as to other qualities. They studied the thermometric characters of the following soils, viz.: Garden earth of dark gray color,—a mixture of sand and gravel with about five per cent of humus; a grayish-white quartz sand; a grayish-brown granite sand; a fine light-gray clay (pipe clay); a yellow sandy clay; and, finally, four lime soils of different physical qualities.

It was found that when the exposure was alike, the dark-gray granite sand became the warmest, and next to this the grayish-white quartz sand. The latter, notwithstanding its lighter color, often acquired a higher temperature at a depth of four inches than the former, a fact to be ascribed to its better conducting power. *The black soils never became so warm as the two just mentioned.* After the black soils, the others came in the following order: garden soil; yellow sandy clay; pipe clay; lime soils having crystalline grains; and, lastly, a pulverulent chalk soil.

To show what different degrees of warmth soils may acquire, under the same circumstances, the following maximum temperatures may be adduced: At noon of a July day, when the temperature of the air was 90°, a thermometer placed at a depth of a little more than one inch, gave these results:

In quartz sand	126°
In crystalline lime soil	115°
In garden soil	114°
In yellow sandy clay	100°
In pipe clay	94°
In chalk soil	87°

Here we observe a difference of nearly 40° in the noonday temperature of the coarse quartz and the chalk soil. Malaguti and Durocher found that the temperature of the garden soil, just below the surface, was, on the average of day and night together, 6° Fahrenheit higher than that of the air, but that this higher temperature diminished at a greater depth. A thermometer buried four inches indicated a mean temperature only 3° above that of the atmosphere.

The experimenters do not mention the influence of water in affecting these results; they do not state the degree of dryness of these soils. It will be seen, however, that the warmest soils are those that retain least water, and doubtless something of the slowness with which the fine soils increase in warmth is connected with the fact that they retain much water, which, in evaporating, appropriates and renders latent a large quantity of heat.

The chalk soil is seen to be the coolest of all, its temperature in these observations being three degrees lower than that of the atmosphere at noonday. In hot climates this coolness is sometimes of great advantage, as appears to happen in Spain, near Cadiz, where the Sherry vineyards flourish. "The Don said the Sherry wine district was very small, not more than twelve miles square. The Sherry grape grew only on certain low, chalky hills, where the earth being light-colored, is not so much burnt; did not chap and split so much by the sun as darker and heavier soils do. A mile beyond these hills the grape deteriorates."—(Dickens' *Household Words* Nov. 13, 1858.)

In Explanation of these observations we must recall to mind the fact that all bodies are capable of absorbing and radiating as well as reflecting heat. These properties, although never dissociated from color, are not necessarily dependent upon it. They chiefly depend upon the character of the surface of bodies. Smooth, polished surfaces absorb and radiate heat least readily; they reflect it most perfectly. Radiation and absorption are opposed to each other, and the power of any body to radiate, is precisely equal to its faculty of absorbing heat.

It must be understood, however, that bodies may differ in their power of absorbing or radiating *heat of different degrees of intensity.* Lamp-black absorbs and radiates heat of all intensities in the same degree. White-lead absorbs heat of low intensity (such as radiates from a vessel filled with boiling water) as fully as lamp-black, but of the intense heat of a lamp it absorbs only about one-half as much. Snow seems to resemble white-lead in this respect. If a black cloth or black paper be spread on the surface of snow, upon which the sun is shining, it will melt much faster under the cloth than elsewhere, and this, too, if the cloth be not in contact with, but suspended above, the snow. In our latitude every one has had opportunity to observe that snow thaws most rapidly when covered by or lying on black earth. The people of Chamouni, in the Swiss Alps, strew the surface of their fields with black-slate powder to hasten the melting of the snow. The reason is that snow absorbs heat of low intensity with greatest facility. The heat of the sun is converted from a high to a low intensity by being absorbed and then radiated by the black material. But it is not color that determines this difference of absorptive power, for indigo and Prussian blue, though of nearly the same color, have very different absorptive powers. So far, however, as our observations extend, it appears that, usually, dark-colored soils absorb heat most rapidly, and that the sun's rays have least effect on light-colored soils. (See the table on p. 196.)

The Rapidity of Change of Temperature independently of color or moisture has been determined on a number of soils by Schübler. A given volume of dry soil was heated to 145°, a thermometer was placed in it, and the time was observed which it required to cool down to 70°, the temperature of the atmosphere being 61°. The subjoined table gives his results. In one column are stated the *times of cooling*, in another the *relative power of retaining heat* or *capacity for heat*, that of lime sand being assumed as 100.

Lime sand	3 hours	30 min		100
Quartz sand	3 "	27 "		95.6
Potter's clay	2 "	41 "		76.9
Gypsum	2 "	34 "		73.8
Clay loam	2 "	30 "		71.8
Clay plow land	2 "	27 "		70.1
Heavy clay	2 "	24 "		68.4
Pure gray clay	2 "	19 "		66.7
Garden earth	2 "	16 "		64.8
Fine carb. lime	2 "	10 "		61.3
Humus	1 "	43 "		49.0
Magnesia	1 "	20 "		38.0

It is seen that the sandy soils cool most slowly, then follow clays and heavy soils, and lastly comes humus.

The order of cooling above given is in all respects identical with that of warming, provided the circumstances are alike. In other words these soils, containing no moisture, or but little, and exposed to heat of low intensity, would be raised through a given range of temperature in the same relative times that they fall through a given number of degrees.

It is to be particularly noticed that dark humus and white magnesia are very closely alike in their rate of cooling, and cool rapidly; while white lime sand stands at the opposite extreme, requiring twice as long to cool to the same extent. These facts strikingly illustrate the great difference between the absorption of radiant heat of low intensity or its communication by conduction on one hand, and that of high intensity like the heat of the sun on the other.

Retention of Heat.—Other circumstances being equal, the power of retaining heat (slowness of cooling) is the greater, the greater the weight of a given bulk of soil, i. e., the larger and denser its particles.

A soil covered with *gravel* cools much more slowly than a sandy surface, and the heat which it collects during a sunny day it carries farther into the night; hence gravelly soils are adapted for such crops as are liable to fail of ripening in cool situations, especially grapes, as has been abundantly observed in practice.

Color is without influence on the loss of heat from the soil by radiation, because the heat is of low intensity. The porosity or roughness of the surface (extent of surface) determines cooling from this cause. Dew, which is deposited as the result of cooling by radiation of heat into the sky, forms abundantly on grass and growing vegetation, and on vegetable mould, but is more rarely met with on coarse sand or gravel.

Influence of Moisture on the Temperature of the Soil.—All soils, when thoroughly wet, seem to be nearly alike in their power of absorbing and retaining warmth. This is due to the fact that the *capacity of water for heat* is much greater than that of the soil. We have seen that lime sand and quartz sand are the slowest of all the ingredients of soils to suffer changes of temperature when exposed to a given source of heat. (See table, p. 194.)

Now, water is nine times slower than quartz in being affected by changes of temperature, and as the entire surface of the wet soil is water, which is, besides, a nearly perfect non-conductor of heat, we can understand that external warmth must affect it slowly.

Again, *the immense consumption of heat in the formation of vapor* (see note, p. 188) must prevent the wet soil from ever acquiring the temperature it shortly attains when dry.

From this cause the difference in temperature between dry and wet soil may often amount to from 10° to 18°.

On this point, again, Schübler furnishes us with the results of his experiments. Columns 4 and 5 in the table below give the temperatures which the thermometer attained when its bulb was immersed in various soils, both wet and dry, each having its natural color. (Columns 1 and 2 are referred to on p. 191.)

	1 Surface.	2	3	4 Surface.	5	6
	Whitened.	Blackened.	Difference.	Wet.	Dry.	Difference.
Magnesia, pure white...............	173.7	121.3	12.6°	95.2	108.7°	13.5°
Fine carbonate of lime, white.......	109.2	122.9	13.7°	96.1	109.4°	13.3°
Gypsum, bright white-gray..........	110.3	124.3	14.0°	97.3°	110.5°	13.2°
Plow land, gray...................	107.6	122.0	11.4°	97.7	111.7°	14.0°
Sandy clay, yellowish..............	108.3°	121.6°	13.3°	98.2°	111.4°	13.2°
Quartz sand, bright yellowish-gray...	109.9	123.6	13.7°	99.1	112.6°	13.5°
Loam, yellowish...................	107.8°	121.1°	13.3°	99.1	112.1°	13.0°
Lime sand, whitish-gray............	109.9	124.0°	14.1°	99.3°	112.1°	12.8°
Heavy clay soil, yellowish-gray.....	107.4°	120.4°	13.0°	99.3	112.3°	13.0°
Pure clay, bluish-gray..............	106.3°	120.0°	13.7°	99.5	113.0	13.5°
Garden mould, blackish-gray........	108.3	122.5°	14.2°	99.5	113.5°	14.0°
Slaty marl, brownish-red...........	108.3°	123.4°	15.1	101.8	115.3	13.5°
Humus, brownish-black............	108.5	120.9°	12.4°	103.6°	117.3°	13.7°

We note that the difference in favor of the dry earth is almost uniformly 13° to 14°. This difference is the same as observed between the whitened and blackened specimens of the same soils. (Column 3.)

We observe, however, that the wet soil in no case becomes as warm as the same soil whitened. We notice further that of the wet soils, the dark-colored ones, humus and marl, are most highly heated. Further it is seen that coarse lime sand (carbonate of lime) acquires 3° higher temperature than fine carbonate of lime, both wet, probably because evaporation proceeded more slowly from the coarse than from the fine materials. Again it is plain on comparing columns 1, 2, and 5, that the gray to yellowish brown and black colors of all the soils, save the first three, assist the elevation of temperature, which rises nearly with the deepening of the color, until in case of humus it lacks but a few degrees of reaching the warmth of a surface of lamp-black.

According to the observations of Dickinson, made at Abbot's Hill, Hertfordshire, England, and continued through eight years, 90 per cent of the water falling between April 1st and October 1st evaporates from the surface of the soil, only 10 per cent finding its way into drains laid three and four feet deep. The total quantity of water that fell during this time amounted to about 2,900,000 lbs. per acre; of this more than 2,600,000 evaporated from the surface. It has been calculated that to evaporate artificially this enormous mass of water, more than seventy-five tons of coal must be consumed.

Thorough draining, by loosening the soil and causing a rapid removal from below of the surplus water, has a most decided influence, especially in spring time, in warming the soil and bringing it into a suitable condition for the support of vegetation.

It is plain, then, that even if we knew with accuracy what are the physical characters of a surface soil, and if we were able to estimate correctly the influence of these characters on its fertility, still we must investigate those circumstances which affect its wetness or dryness, whether they be an impervious subsoil, or springs coming to the surface, or the amount and frequency of rain-falls, taken in connection with other meteorological causes. We cannot decide that a clay is too wet or a sand too dry, until we know its situation and the climate it is subjected to.

The great deserts of the globe do not owe their barrenness to necessary poverty of soil, but to meteorological influences—to the continued prevalence of parching winds, and the absence of mountains, to condense the atmospheric water and establish a system of rivers and streams. This is not the place to enter into a discussion of the causes that may determine or modify climate; but to illustrate the effect that may be produced by means within human control, it may be stated that previous to the year 1821, the French district Provence was a fertile and well-watered region. In 1822, the olive trees which were largely cultivated there were injured by frost, and the inhabitants began to cut them up root and branch. This amounted to clearing off a forest, and, in consequence, the streams dried up, and the productiveness of the country was seriously diminished.

The Angle at which the Sun's Rays Strike a Soil is of great influence on its temperature. The more this approaches a right angle the greater the heating effect. In the latitude of England the sun's heat acts most powerfully on surfaces having a southern exposure, and which are inclined at an angle of 25° and 30°. The best vineyards of the Rhine and Neckar are also on hill-sides, so situated. In Lapland and Spitzbergen the southern side of hills may be seen covered with vegetation, while lasting or even perpetual snow lies on their northern inclinations.

The Influence of a Wall or other Reflecting Surface upon the warmth of a soil lying to the south of it was observed in the case of garden soil by Malaguti and Durocher. The highest temperature indicated by a thermometer placed in this soil at a distance of six inches from the wall, during a series of observations lasting seven days (April, 1852), was 32° Fahrenheit higher at the surface, and 18° higher at a depth of four inches than in the same soil on the north side of the wall. The average temperature of the former during this time was 8° higher than that of the latter. In another trial in March the difference in average temperature between the southern and northern exposures was nearly double this amount in favor of the former.

As is well known, fruits which refuse to ripen in cold climates under ordinary conditions of exposure may attain perfection when trained against the sunny side of a wall. It is thus that in the north of England pears and plums are raised in the most unfavorable seasons, and that the vineyards of Fontainebleau produce such delicious Chasselas grapes for the Paris market, the vines being trained against walls on the Thomery system.

In the Rhine district grape vines are kept low and as near the soil as possible, so that the heat of the sun may be reflected back upon them from the ground, and the ripening is then carried through the nights by the heat radiated from the earth.—(*Journal Highland and Agricultural Society*, July, 1858, p. 347.)

Vegetation.—Malaguti and Durocher also studied the effect of a *sod* on the temperature of the soil. They observed that it hindered the warming of the soil, and indeed to about the same extent as a layer of earth of three inches depth. Thus a thermometer four inches deep in green-sward acquires the same temperature as one seven inches deep in the same soil not grassed.

CHAPTER V.

THE SOIL AS A SOURCE OF FOOD TO CROPS.—INGREDIENTS WHOSE ELEMENTS ARE OF ATMOSPHERIC ORIGIN.

§ 1.

THE FREE WATER OF THE SOIL IN ITS RELATIONS TO VEGETABLE NUTRITION.

Water may exist free in the soil in three conditions, which we designate respectively *hydrostatic, capillary,* and *hygroscopic.*

Hydrostatic or Flowing* Water is water visible as

---

* I. e., capable of flowing.

such to the eye, and free to obey the laws of gravity and motion. When the soil is saturated by rains, melting snows, or by overflow of streams, its pores contain hydrostatic water, which sooner or later sinks away into the subsoil or escapes into drains, streams, or lower situations.

Bottom Water is *permanent hydrostatic water*, reached nearly always in excavating deep soils. The surface of water in a well corresponds with, or is somewhat below, the upper limit of bottom water. It usually fluctuates in level, rising nearer the surface of the soil in wet seasons, and receding during drought. In general, agricultural plants are injured if their roots be immersed for any length of time in hydrostatic water; and soils in which bottom water is found at a little depth during the season of growth are unprofitable for culture.

If this depth be but a few inches, we have a bog, swamp, or swale. If it is one and a half to three feet, and the surface soil be light, gravelly, or open, so as to admit of rapid evaporation, some plants, especially grasses, may flourish. If at a constant depth of four to eight feet under a gravelly or light loamy soil, it is favorable to crops as an abundant source of water.

Heavy clays, which retain hydrostatic water for a long time, being but little permeable, are for the same reasons unfavorable to most crops, unless artificial provision be made for removing the excess.

Rice, as we have seen, (II. C. G., p. 252), is a plant which grows well with its roots situated in water. Henrici's experiment with the raspberry (II. C. G., p. 254), and the frequent finding of roots of clover, turnips, etc., in cisterns or drain pipes, indicate that many or all agricultural plants may send down roots into the bottom water for the purpose of gathering a sufficient supply of this necessary liquid.

Capillary Water is that which is held in the fine pores of the soil by the surface attraction of its particles, as oil is held in the wick of a lamp. The adhesion of the water to the particles of earth suspends the flow of the liquid, and it is no longer subject to the laws of hydrostatics. Capillary water is usually designated as *moisture*, though a soil saturated with capillary water would be, in most cases, *wet*. The capillary power of various soils has already been noticed, and is for coarse sands $25°|_0$; for loams and clays, 40 to $70°|_0$; for garden mould and humus, much higher, 80 to $300°|_0$. (See p. 180.)

For a certain distance above bottom water, the soil is saturated with capillary water, and this distance is the greater, the greater the capillary power of the soil, i. e., the finer its pores.

Capillary water is not visible as a distinct liquid layer on or between the particles of soil, but is still recognizable by the eye. Even in the driest weather and in the driest sand (that is, when not shut off from bottom water by too great distance or an intervening gravelly subsoil) it may be found one or a few inches below the surface where the soil *looks moist*—has a darker shade of color.

Hygroscopic Water is that which is not perceptible to the senses, but is appreciated by loss or gain of weight in the body which acquires or is deprived of it. (H. C. G., p. 54.) The loss experienced by an air-dry soil when kept for some hours at, or slightly above, the boiling point (212° F.,) expresses its content of hygroscopic water. This quantity is variable according to the character of the soil, and is constantly varying with the temperature; increasing during the night when it is collected from the atmosphere, and diminishing during the day when it returns in part to the air. (See p. 164.) The amount of hygroscopic water ranges from 0.5 to 10 or more per cent.

Value of these Distinctions.—These distinctions between hydrostatic, capillary, and hygroscopic water, are nothing absolute, but rather those of degree. Hygroscopic water is capillary in all respects, save that its quantity is small, and its adhesion to the particles of soil more firm for that reason. Again, no precise boundary can always be drawn between capillary and hydrostatic water, especially in soil having fine pores. The terms are nevertheless useful in conveying an idea of the degrees of wetness or moisture in the soil.

Roots Absorb Capillary or Hygroscopic Water.—It is from capillary or hygroscopic water that the roots of most agricultural plants chiefly draw a supply of this liquid, though not infrequently they send roots into wells and drains. The physical characters of soils that have been already considered suffice to explain how the *earth* acquires this water; it here remains to notice how the plant is related to it.

As we have seen (pp. 35–38), the aerial organs appear incapable of taking up either vapor or liquid water from the air to much extent, and even roots continually exhale vapor without absorbing any, or at least without being able to make up the loss which they continually suffer.

Transpiration of Water through Plants.—It is a most familiar fact that water constantly exhales from the surface of the plant. The amount of this exhalation is often very great. Hales, the earliest observer of this phenomenon, found that a sunflower whose foliage had 39 square feet of surface, gave off in 24 hours 3 lbs. of water. A cabbage, whose surface of leaves equaled 19 square feet, exhaled in the same time very nearly as much. Schleiden found the loss of water from a square foot of grass-sod to be more than $1\frac{1}{2}$ lbs. in 24 hours. Schübler states that in the same time 1 square foot of pasture-grass exhaled nearly $5\frac{1}{2}$ lbs. of water. In one of Knop's more recent experiments, (*Vs. St.*, VI, 239), a dwarf bean exhaled during 23 days, in September and October, 13 times its weight of water. In another trial a maize-plant transpired 36 times its weight of water, from May 22d to Sept. 4th. According to Knop, a grass-plant will exhale its own weight of water in 24 hours of hot and dry summer weather.

The water exhaled from the leaves must be constantly supplied by *absorption* at the roots, else the foliage soon becomes flabby or wilts, and finally dies. Except so far as water is actually formed or fixed within the plant, its absorption at the roots, its passage through the tissues, and its exhalation from the foliage, are nearly equal in quantity and mutually dependent during the healthy existence of vegetation.

Circumstances that Influence Transpiration.—*a. The structure of the leaf*, including the character of the epidermis, and the number of *stomata* as they affect exhalation, has been considered in "How Crops Grow," (pp. 286-8).

b. The physical conditions which facilitate evaporation increase the amount of water that passes through the plant. Exhalation of water-vapor proceeds most rapidly in a hot, dry, windy summer day. It is nearly checked when the air is saturated with moisture, and varies through a wide range according to the conditions just named.

c. The oxidations that are constantly going on within the plant may, under certain conditions, acquire sufficient intensity to develop a perceptible amount of heat and cause the vaporization of water. It has been repeatedly noticed that the process of flowering is accompanied by considerable elevation of temperature, (p. 24). In general, however, the opposite process of deoxidation preponderates with the plant, and this must occasion a reduction of temperature. These interior changes can have no appreciable influence upon transpiration as compared with those that depend upon external causes. Sachs found in some of his experiments (p. 36) that exhalation took place from plants confined in a limited space over water. Sachs believed that the air surrounding the plants in these experiments was saturated with vapor of water, and concluded that heat was developed within the plant, which caused vaporization. More recently, Bochm (*Sitzungsberichte der Wiener Akad.*, XLVIII, 15) has made probable that the air was not fully or constantly saturated with moisture in these experiments, and by taking greater precautions has arrived at the conclusion that transpiration absolutely ceases in air saturated with aqueous vapor.

d. The condition of the tissues of the plant, as dependent upon their age and vegetative activity, likewise has a marked effect on transpiration. Lawes[*] and Knop both found that young plants lose more water than older ones. This is due to the diminished power of mature foliage to imbibe and contain water, its cells becoming choked up with growth and inactive.

e. The character of the medium in which the roots are situated also remarkably influences the rate of transpiration. This fact, first observed by Mr. Lawes, in 1850, *loc. cit.*, was more distinctly brought out by Dr. Sachs at a later period. (*Vs. St.*, I, p. 203.)

Sachs experimented on various plants, viz.: beans, squashes, tobacco, and maize, and observed their transpiration in weak solutions (mostly containing one per cent) of nitre, common salt, gypsum, (one-fifth per cent solution) and sulphate of ammonia. He also experimented with maize in a mixed solution of phosphate and silicate of potash, sulphates of lime and magnesia, and common salt, and likewise observed the effect of free nitric acid and free potash on the squash plant. The young plants were either germinated in the soil, then removed from it and set with their rootlets in the solution, or else were kept in the soil and watered with the solution. The glass

---

[*] *Experimental Investigation into the Amount of Water given off by Plants during their Growth*, by J. B. Lawes, of Rothamstead, London, 1850.

vessel containing the plant and solution was closed above, around the stem of the plant, by glass plates and cement, so that no loss of water could occur except through the plant itself, and this loss was ascertained by daily weighings. The result was that all the solutions mentioned, except that of free nitric acid, quite uniformly retarded transpiration to a degree varying from 10 to 90 per cent, while the free acid accelerated the transpiration in a corresponding manner.

Sachs experimented also with four tobacco plants, two situated in coarse sand and two in yellow loam. The plants stood side by side exposed to the same temperature, etc., and daily weighings were made during a week or more, to learn the amount of exhalation. The result was that the total loss, as well as the daily loss in the majority of weighings, was greater from the plant growing in loam, although through certain short periods the opposite was noticed.

f. The temperature of the soil considerably affects the rate of transpiration by influencing the amount of absorption at the roots. Sachs made a number of weighings upon two tobacco plants of equal size, potted in portions of the same soil and having their foliage exposed to the same atmosphere. After observing their relative transpiration when their roots were at the same temperature, one pot was warmed a number of degrees, and the result was invariably observed that elevating the temperature of the soil increased the transpiration.

The same observer subsequently noticed the entire suppression of absorption by a reduction of temperature to 41° to 43° F. A number of healthy tobacco and squash plants, rooted in a soil kept nearly saturated with water, were growing late in November in a room, the temperature of which fell at night to the point just named. In the morning the leaves of these plants were so wilted that they hung down like wet cloths, as if the soil were completely dry, or they had been for a long time acted upon by a powerful sun. Since, however, the soil was moist, the wilting could only arise from the inability of the roots to absorb water as rapidly as it exhaled from the leaves, owing to the low temperature. Further experiments showed that warming the soil in which the wilted plants stood, restored the foliage to its proper turgidity in a short time, and by surrounding the soil of a fresh plant with snow, the leaves wilted in three or four hours.

Cabbages, winter colza, and beans, similarly circumstanced, did not wilt, showing that different plants are unequally affected. The general rule nevertheless appears to be established that within certain limits the root absorbs more vigorously at high than at low temperatures.

The Amount of Loss of Water of Vegetation in Wilting has been determined by Hesse (*Vs. St.*, I, 248) in case of sugar-beet leaves. Of two similar leaves, one, gathered at evening after several days of dryness and sunshine, contained $85.74°|_0$ of water; the other, gathered the next morning, two hours after a rain storm, yielded $89.57°|_0$. The difference was accordingly $3.8°|_0$. Other observations corroborated this result.

Is Exhalation Indispensable to Plants?—It was for a long time supposed that transpiration is indispensable to the life of plants. It was taught that the water which the plant imbibes from the soil to replace that lost by exhalation, is the means of bringing into its roots the mineral and other soluble substances that serve for its nutriment.

There are, however, strong grounds for believing that the current of water which ascends through a plant moves independently of the matters that may be in solution, either without or within it; and, moreover, the motion of soluble matters from the soil into the plant may go on, although there be no ascending aqueous current. (II. C. G., pp. 288 and 340.)

In accordance with these views, vegetation grows as well in the confined atmosphere of green-houses or of Wardian Cases, where the air is for the most part or entirely saturated with vapor, so that transpiration is reduced to a minimum, as in the free air, where it may attain a maximum. As is well known, the growth of field crops and garden vegetables is often most rapid during damp and showery weather, when the transpiration must proceed with comparative slowness.

While the above considerations, together with the assertion of Knop, that leaves lose for the first half hour nearly the same quantities of water under similar exposure, whether they are attached to the stem or removed from it, whether entire or in fragments, would lead to the conclusion that transpiration, which is so extremely variable in its amount, is, so to speak, an accident to the plant and not a process essential to its existence or welfare, there are, on the other hand, facts which appear to indicate the contrary.

In certain experiments of Sachs, in which the roots of a bean were situated in an atmosphere nearly saturated with aqueous vapor, the foliage being exposed to the air, although the plant continued for two months fresh and healthy to appearance, it remained entirely stationary in its development. (*Vs. St.*, I, 237.)

Knop also mentions incidentally (*Vs. St.*, I, 192) that beans, lupines, and maize, die when the whole plant is kept confined in a vessel over water.

It is not, however, improbable that the cessation of growth in the one case and the death of the plants in the other were due not so much to the checking of transpiration, which, as we have seen, is never entirely suppressed under these circumstances, as to the exhaustion of oxygen or the undue accumulation of carbonic acid in the narrow and confined atmosphere in which these results were noticed.

On the whole, then, we conclude from the evidence before us that transpiration is not necessary to vegetation, or at least fulfills no very important offices in the nutrition of plants.

The entrance of water into the plant and the steady maintenance of its proper content of this substance, under all circumstances is of the utmost moment, and leads us to notice in the next place the

Direct Proof that Crops can Absorb from the Soil enough Hygroscopic Water to Maintain their Life.—Sachs suffered a young bean-plant standing in a pot of very retentive (clay) soil to remain without watering until the leaves began to wilt. A high and spacious glass cylinder, having a layer of water at its bottom, was then provided, and the pot containing the wilting plant was supported in it, near its top, while the cylinder was capped by two semicircular plates of glass which closed snugly about the stem of the bean. The pot of soil and the roots of the plant were thus enclosed in an atmosphere which was constantly saturated, or nearly so, with watery vapor, while the leaves were fully exposed to the free air. It was now to be observed whether the water that exhaled from the leaves could be supplied by the hygroscopic moisture which the soil should gather from the damp air enveloping it. This proved to be the case. The leaves, previously wilted, recovered their proper turgidity, and remained fresh during the two months of June and July.

Sachs, having shown in other experiments that plants situated precisely like this bean, save that the roots are not in contact with soil, lose water continuously and have no power to recover it from damp air (p. 36) thus gives us demonstration that the clay soil which condenses vapor in its pores and holds it as hygroscopic water, yields it again to the plant, and thus becomes the medium through which water is continually carried from the atmosphere into vegetation.

In a similar experiment, a tobacco plant was employed which stood in a soil of humus. This material was also capable of supplying the plant with water by virtue of its hygroscopic power, but less satisfactorily than the clay. As already mentioned, these plants, while remaining fresh, exhibited no signs of growth. This may be due to the consumption of oxygen by the roots and soil, or possibly the roots of plants may require an occasional drenching with liquid water. Further investigations in this direction are required and promise most interesting results.

What Proportion of the Capillary and Hygroscopic Water of the Soil may Plants Absorb, is a question that Dr. Sachs has made the only attempts to answer. When a plant, whose leaves are in a very moist atmosphere, wilts or begins to wilt in the night time, when therefore transpiration is reduced to a minimum, it is because the soil no longer yields it water. The quantity of water still contained in a soil at that juncture is that which the plant cannot remove from it,—is that which is unavailable to vegetation, or at least to the kind of vegetation experimented with. Sachs made trials on this principle with tobacco plants in three different soils.

The plant began to wilt in a mixture of *black humus* (from beech-wood) and *sand*, when the soil contained $12.3°|_0$ of water.* This soil, however, was capable of holding $46°|_0$ of capillary water. It results therefore that of its highest content of absorbed water $33.7°|_0$ $(=46-12.3)$ was available to the tobacco plant.

Another plant began to wilt on a rainy night, while the *loam* it stood in contained $8°|_0$ of water. This soil was able to absorb $52.1°|_0$ of water, so that it might after

---

* Ascertained by drying at 212°.

saturation, furnish the tobacco plant with $44.1°|_0$ of its weight of water.

A *coarse sand* that could hold $20.8°|_0$ of water was found to yield all but $1.5°|_0$ to a tobacco plant.

From these trials we gather with at least approximate accuracy the power of the plant to extract water from these several soils, and by difference, the quantity of water in them that was unavailable to the tobacco plant.

How do the Roots take Hygroscopic Water from the Soil?—The entire plant, when living, is itself extremely hygroscopic. Even the dead plant retains a certain proportion of water with great obstinacy. Thus wheat, maize, starch, straw, and most air-dry vegetable substances, contain 12 to $15°|_0$ of water; and when these matters are exposed to damp air, they can take up much more. According to Trommer (*Bodenkunde*, p. 270), 100 parts of the following matters, when dry, absorb from moist air in

	12	24	48	72	
	\multicolumn{4}{c}{hours.}	
Fine cut barley straw,	15	24	34	45	parts of water.
" " rye "	12	20	27	29	" " "
" " white unsized paper,	8	12	17	19	" " "

As already explained, a body is hygroscopic because there is *attraction* between its particles and the particles of water. The form of attraction exerted thus among different kinds of matter is termed adhesive attraction, or simply adhesion.

Adhesion acts only through a small distance, but its intensity varies greatly within this distance. If we attempt to remove hygroscopic water from starch or any similar body by drying at 212°, we shall find that the greater part of the moisture is easily expelled in a short time, but we shall also notice that it requires a relatively much longer time to expel the last portions. A general law of attraction is that its force diminishes as the distance between the attracting bodies increases. This has been exactly demonstrated in case of the force of gravity and electrical attraction, which act through great intervals of space.

We must therefore suppose that when a mass of hygroscopic matter is allowed to coat itself with water by the exercise of its adhesive attraction, the layer of aqueous particles which is in nearest contact is more strongly held to it than the next outer layer, and the adhesion diminishes with the distance, until, at a certain point, still too small for us to perceive, the attraction is nothing, or is neutralized by other opposing forces, and further adhesion ceases.

Suppose, now, we bring in contact at a single point two masses of the same kind of matter, one of which is saturated with hygroscopic water and the other is perfectly dry. It is plain that the outer layers of water-particles adhering to the moist body come at once within the range of a more powerful attraction exerted by the very surface of the dry body. The external particles of water attached to the first must then pass to the second, and they must also distribute themselves equally over the surface of the latter; and this motion must go on until the attraction of the two surfaces is equally satisfied, and the water is equally distributed according to the surface, i. e., is uniform over the whole surface.

If of two different bodies put in contact (one dry and one moist) the surfaces be equal, but the attractive force of one for water be twice that of the other, then motion must go on until the one has appropriated two-thirds, and the other is left with one-third the total amount of water.

When bodies in contact have thus equalized the water at their disposal, they may be said to be in a condition of *hygroscopic equilibrium*. Any cause which disturbs this equilibrium at once sets up motion of the hygroscopic water, which always proceeds from the more dry to the less dry body.

The application of these principles to the question before us is apparent. The young, active roots that are in contact with the soil are eminently hygroscopic, as is demonstrated by the fact that they supply the plant with large quantities of water when the soil is so dry that it has no visible moisture. They therefore share with the soil the moisture which the latter contains. As water evaporates from the surface of the foliage, its place is supplied by the adjacent portions, and thus motion is established within the plant which propagates itself to the roots and through these to the soil.

Each particle of water that flies off in vapor from the leaf makes room for the entrance of a particle at the root. If the soil and air have a surplus of water, the plant will contain more; if the soil and air be dry, it will contain less. Within certain narrow limits the supply and waste may vary without detriment to the plant, but when the loss goes on more rapidly than the supply can be kept up, or when the absolute content of water in the soil is reduced to a certain point, the plant shortly wilts. Even then its content of water is many times greater than that of the soil. The living tobacco plant cannot contain less than $80°|_0$ of water, while the soils in Sachs' experiments contained but $12.3°|_0$ and $1.5°|_0$ respectively. When fully air-dry, vegetable matter retains $13°|_0$ to $15°|_0$ of water, while the soil similarly dry rarely contains more than $1-2°|_0$.

The plant therefore, especially when living, is much more hygroscopic than the soil.

If roots are so hygroscopic, why, it may be asked, do they not directly absorb vapor of water from the air of the soil? It cannot be denied that both the roots and foliage of plants are capable of this kind of absorption, and that it is taking place constantly in case of the roots. The experiments before described prove, however, that the higher orders of plants absorb *very little* in this way.

too little, in fact, to be estimated by the methods hitherto employed. Sachs explains this as follows: Assuming that the roots have at a given temperature as strong an attraction for water in the state of vapor as for liquid water, the amount of each taken up in a given time under the same circumstances would be in proportion to the weight of each contained in a given space. A cubic inch of water yields at 212 nearly a cubic foot (accurately, 1,696 times its volume, the barometer standing at 29.92 inches) of vapor. We may then assume that the absorption of liquid or hygroscopic water proceeds at least one thousand times more rapidly than that of vapor, a difference in rate that enables us to comprehend why a plant may gain water by its roots from the soil, when it would lose water by its roots were they simply stationed in air saturated with vapor.

Again, the soil need not be more hygroscopic than roots, to supply the latter with water. It is important only that it present a *sufficient surface*. As is well known, a plant requires a great volume of earth to nourish it properly, and the root-surface is trifling, compared to the surface of the particles which compose the soil.

Boussingault found by actual measurement that, according to the rules of garden culture as practiced near Strasburg, a dwarf bean had at its disposition 57 pounds of soil; a potato plant, 190 pounds; a tobacco plant, 470 pounds; and a hop plant, 2,900 pounds. These weights correspond to about 1, 3, 7, and 50 cubic feet respectively.

The Quantity of Water in Vegetation is influenced by that of the Soil.—De Saussure observed that plants growing in a dry lime soil contained less water than those from a loam. It is well known that the grass of a wet summer is taller and more succulent, and the green crop is heavier than that from the same field in a dry summer. It does not, however, make much more hay, its greater weight consisting to a large degree of water, which is lost in drying. Ritthausen gives some data concerning two clover crops of the year 1854, from a loamy sand, portions of which were manured, one with ashes, others with gypsum.

The following statement gives the produce of the nearly* fresh and of the air-dry crops.

	Weight in pounds per acre.		
	Fresh.	Air-dry.	Water lost in drying.
Crop I, manured with ashes,	14,903	5,182	9,721
" " unmanured,	12,380	5,418	6,962
Crop II. manured with gypsum,	22,256	4,800	17,456
" " unmanured,	18,815	5,190	13,625

It is seen that while in both cases the fresh manured crop greatly outweighed the unmanured, the excess of weight consisted of *water*. In fact, the unmanured plots yielded *more hay* than the manured. The manured clover was darker in color than the other, and the stems were large and hollow, i. e., by rapid growth the pith cells were broken away from each other and formed only a lining to the stalk, while in the unmanured clover the pith remained undisturbed, the stems being more compact in structure. (II. C. G., p. 369.)

The Quantity of Soil-water most favorable to Crops has been studied by Ilienkoff and Hellriegel. The former (*Ann. der. Chem. u. Ph.* 136, p. 160,) experimented with buckwheat plants stationed in pots filled with garden earth. The pots were of the same size and had the same exposure at the south side of an apartment. The plants received at each watering in

Pot No. 1, $\frac{1}{2}$ liter of water
" " 2, $\frac{1}{4}$ " "
" " 3, $\frac{1}{8}$ " "
" " 4, $\frac{1}{16}$ " "
" " 5, $\frac{1}{32}$ " "

---

* The clover was collected from the surface of a Saxon square ell, and was somewhat wilted before coming into Ritthausen's hands. The quantities above given are calculated to English acres and pounds.

The waterings were made simultaneously at the moment when all the water previously given to No. 1 was absorbed by the soil. During the 67 days of the experiment the plants were watered 17 times. The subjoined table gives the results:

No. of pot.	Weight of fresh Crops in grams.	Weight of dry Crops in grams.		Number of Seeds.	Liters of water used.
		STRAW.	SEEDS.		
1	27.99	4.52	1.68	111	25.0
2	65.05	8.47	5.47	283	12.5
3	24.95	4.55	1.73	93	6.25
4	9.98	1.41	0.52	37	3.12
5	2.30	0.30	0.09	12	1.56

The experiment demonstrates that the quantity of water supplied to a plant has a decided effect upon the yield. Pot No. 2 was most favorably situated in this respect. No. 1 had a surplus of water and the other pots received too little. The experiment does not teach what proportion of water in the soil was most advantageous, for neither the weight of the soil nor the size of the pot is mentioned.

Hellriegel (*Chem. Ackersmann*, 1868, p. 15) experimented with wheat, rye, and oats, in a pure sand mixed with a sufficiency of plant-food. The sand when saturated with water contained 25°|₀ of the liquid. The following table gives further particulars of his experiments and the results. The weights are grams.

WATER IN THE SOIL.		YIELD OF WHEAT.		YIELD OF RYE.		YIELD OF OATS.	
In per cent of Soil.	In per cent of retentive power.	Straw and Chaff.	Grain.	Straw and Chaff.	Grain.	Straw and Chaff.	Grain.
2½–5	10–20	7.0	2.8	8.3	3.9	4.2	1.8
5 –10	20–40	15.1	8.4	11.8	8.1	11.8	7.8
10 –15	40–60	21.4	10.3	15.1	10.3	13.9	10.9
15 –20	60–80	23.3	11.4	16.4	10.3	15.8	11.8

In each case the proportion of water in the soil was preserved within the limits given in the first column of the table, throughout the entire period of growth. It is seen that in this sandy soil 10–15 per cent of water enabled rye to yield a maximum of grain and brought wheat and oats very closely to a maximum crop. Hellriegel noticed that the plants exhibited no visible symptoms of deficiency of water, except through stunted growth, in any of these experiments. Wilting never took place except when the supply of water was less than $2^1\!/_2$ per cent.

Grouven (*Ueber den Zusammenhang zwischen Witterung, Boden und Düngung in ihrem Einflusse auf die Quantität und Qualität der Erndten*, Glogau, 1868) gives the results of an extensive series of field trials, in which, among other circumstances, the influence of water upon the crops was observed. His discussion of the subject is too detailed to reproduce in this treatise, but the great influence of the supply of water (by rain, etc.,) is most strikingly brought out. The experimental fields were situated in various parts of Germany and Austria, and were cultivated with sugar beets in 1862, under the same fertilizing applications, as regards both kind and quantity. Of 14 trials in which records of the rain-fall were kept, the 8 best crops received from the time of sowing, May 1st, to that of harvesting, Oct. 15th, an average quantity of rain equal to 140 Paris lines in depth. The 6 poorest crops received in the same time on the average but 115 lines. During the most critical period of growth, viz., between the 20th of June and the 10th of September, the 8 best crops enjoyed an average rain-fall of 90.7 lines, while the 6 poorest received but 57.7 lines.

It is a well recognized fact that next to temperature, the water supply is the most influential factor in the product of a crop. Poor soils give good crops in seasons of plentiful and well-distributed rain or when skillfully irrigated, but insufficient moisture in the soil is an evil that no supplies of plant-food can neutralize.

The Functions of Water in the Nourishment of Vegetation, so far as we know them, are of two kinds

In the first place it is an unfailing and sufficient source of its elements,—hydrogen and oxygen,—and undoubtedly enters directly or indirectly into chemical combination with the carbon taken up from carbonic acid, to form sugar, starch, cellulose, and other carbohydrates. In the second place it performs important physical offices; is the vehicle or medium of all the circulation of matters in the plant; is directly concerned, it would appear, in imbibing gaseous food in the foliage and solid nutriment through the roots; and by the force with which it is absorbed, directly influences the enlargement of the cells, and, perhaps, also the direction of their expansion,—an effect shown by the facts just adduced relative to the clover crops examined by Ritthausen.

Indirectly, also, water performs the most important service of continually solving and making accessible to crops the solid matters in the vicinity of their roots, as has been indicated in the chapter on the Origin of Soils.

Combined Water of the Soil.—As already stated, there may exist in the soil compounds of which water is a chemical component. True clay (kaolinite) and the zeolites, as well as the oxides of iron that result from weathering, contain chemically combined water. Hence a soil which has been totally deprived of its hygroscopic water by drying at 212°, may, and, unless consisting of pure sand, does, yield a further small amount of water by exposure to a higher heat. This combined water has no direct influence on the life of the plant or on the character of the soil, except so far as it is related to the properties of the compounds of which it is an ingredient.

§ 2.

THE AIR OF THE SOIL.

As to the free Oxygen and Nitrogen which exist in the interstices or adhere to the particles of the soil, there is little to add here to what has been remarked in previous paragraphs.

Free Oxygen, as De Saussure and Traube have shown, is indispensable to growth, and must therefore be accessible to the roots of plants.

The soil, being eminently porous, condenses oxygen. Blumtritt and Reichardt indeed found no considerable amount of condensed oxygen in most of the soils and substances they examined (p. 167); but the experiments of Stenhouse (p. 169) and the well-known deodorizing effects of the soil upon fecal matters, leave no doubt as to the fact. The condensed oxygen must usually spend itself in chemical action. Its proportion would appear not to be large; but, being replaced as rapidly as it enters into combination, the total quantity absorbed may be considerable. Organic matters and lower oxides are thereby oxidized. Carbon is converted into carbonic acid, hydrogen into water, protoxide of iron into peroxide. The upper portions of the soil are constantly suffering change by the action of free oxygen, so long as any oxidable matters exist in them. These oxidations act to solve the soil and render its elements available to vegetation. (See p. 131.)

Free Nitrogen in the air of the soil is doubtless indifferent to vegetation. The question of its conversion into nitric acid or ammonia will be noticed presently. (See p. 259.)

Carbonic Acid.—The air of the soil is usually richer in carbonic acid, and poorer in oxygen, than the normal atmosphere, while the proportion (by volume) of nitrogen is the same or very nearly so. The proportions of carbonic acid by weight in the air included in a variety of soils have already been stated. Here follow the total quantities of this gas and of air, as well as the composition of the latter in 100 parts by volume, as determined by

Boussingault and Lewy. (*Mémoires de Chimie Agricole, etc.*, p. 369.)

	Cubic feet of air in acre to depth of 14 inches.	Cubic feet of carbonic acid in acre to depth of 14 inches.	Composition of the air in the soil in 100 parts by volume.		
			Carbonic acid.	Oxygen.	Nitrogen.
Sandy subsoil of forest....................	4416	11	0.24	19.66	79.55
Loamy " "	3530	28	0.79	19.66	79.55
Surface soil " "	5891	57	0.87	19.61	79.52
Clayey " of artichoke field	10310	71	0.66	19.99	79.35
Soil of asparagus bed not manured for one year	11182	86	0.74	19.02	80.24
" " " " newly manured.........	11182	172	1.54	18.80	79.66
Sandy soil, six days after manuring.... [of rain	11783	257	2.21		
" " ten " " " three days	11783	1144	9.71	10.35	79.91
Vegetable mold-compost....................	21049	772	3.64	16.45	79.91

	Cubic feet of air over one acre to height of 14 inches.	Cubic feet of carbonic acid in air over one acre to height of 14 inches.	Composition of air above the soil in 100 parts.		
			Carbonic acid.	Oxygen.	Nitrogen.
	50820	12	0.025	20.945	79.030

The percentage, as well as the absolute quantity of carbonic acid, is seen to stand in close relation with the organic matters of the soil. The influence of the recent application of manure rich in organic substances is strikingly shown in case of the asparagus bed and the sandy soil. The lowest percentage of carbonic acid is 10 times that of the atmosphere a few feet above the surface of the earth, as determined at the same time, while the highest percentage is 390 times that proportion.

Even in the sandy subsoil the quantity of free carbonic acid is as great as in an equal bulk of the atmosphere; and in the cultivated soils it is present in from 6 to 95 times greater amount. In other words, in the cultivated soils taken to the depth of 14 inches, there was found as much carbonic acid gas as existed in the same horizontal area of the atmosphere through a height of 7 to 110 feet.

The accumulation of such a percentage of carbonic acid gas in the interstices of the soil demonstrates the rapid formation of this substance, which must as rapidly diffuse off into the air. The roots, and, what is of more significance, the leaves of crops, are thus far more copiously fed with this substance than were they simply bathed by the free atmosphere so long as the latter is unagitated.

When the wind blows, the carbonic acid of the soil is of less account in feeding vegetation compared with that of the atmosphere. When the air moves at the rate of two feet per second, the current is just plainly perceptible. A mass of foliage 2 feet high and 200 feet* long, situated in such a current, would be swept by a volume of atmosphere, amounting in one minute to 48,000 cubic feet, and containing 12 cubic feet of carbonic acid. In one hour it would amount to 2,880,000 cubic feet of air, equal to 720 cubic feet of carbonic acid, and in one day to 69,120,000 cubic feet of air, containing no less than 17,280 cubic feet of carbonic acid.

In a brisk wind, ten times the above quantities of air and carbonic acid would pass by or through the foliage. It is plain, then, that the atmosphere, which is rarely at rest, can supply carbonic acid abundantly to foliage without the concourse of the soil. At the same time it should not be forgotten that the carbonic acid of the atmosphere is largely derived from the soil.

Carbonic Acid in the Water of the Soil.—Notwithstanding the presence of so much carbonic acid in the air of the soil, it appears that the capillary soil-water, or so

---

* A square field containing one acre is 208 feet and a few inches on each side.

much of it as may be expressed by pressure, is not nearly saturated with this gas.

De Saussure (*Recherches Chimiques sur la Végétation*, p. 168) filled large vessels with soils *rich in organic matters*, poured on as much water as the earth could imbibe, allowing the excess to drain off and the vessels to stand five days. Then the soils were subjected to powerful pressure, and the water thus extracted was examined for carbonic acid. It contained but 2% of its volume of the gas.

Since at a medium temperature ($60°$ F.) water is capable of dissolving 100% (its own bulk) of carbonic acid, it would appear on first thought inexplicable that the soil-water should hold but 2 per cent. Henry and Dalton long ago demonstrated that the relative proportion in which the ingredients of a gaseous mixture are absorbed by water depends not only on the relative solubility of each gas by itself, but also on the proportions in which they exist in the mixture. The large quantities of oxygen, and especially of nitrogen, associated with carbonic acid in the pores of the soil, thus act to prevent the last-named gas being taken up in greater amount; for, while carbonic acid is about fifty times more soluble than the atmospheric mixture of oxygen and nitrogen, the latter is present in fifty times (more or less) the quantity of the former.

Absorption of Carbonic Acid by the Soil.—According to Van den Broek, (*Ann. der Chemie u. Ph.*, 115, p. 87) certain wells in the vicinity of Utrecht, Holland, which are excavated only a few feet deep in the soil of gardens, contain water which is destitute of carbonic acid (gives no precipitate with lime-water), while those which penetrate into the underlying sand contain large quantities of carbonate of lime in solution in carbonic acid.

Van den Broek made the following experiments with garden-soil newly manured, and containing free carbonic acid in its interstices, which could be displaced by a current of air. Through a mass of this earth 20 inches deep and 3 inches in diameter, pure distilled water (free from carbonic acid) was allowed to filter. *It ran through without taking up any of the gas.* Again, water containing its own volume of carbonic acid was filtered through a similar body of the same earth. *This water gave up all its carbonic acid while in contact with the soil.* After a certain amount had run off, however, the subsequent portions contained it. In other words, the soils experimented with were able to absorb carbonic acid from its aqueous solution, even when their interstices contained the gas in the free state. These extraordinary phenomena deserve further study.

§ 3.

NON-NITROGENOUS ORGANIC MATTERS OF THE SOIL.—
CARBOHYDRATES. VEGETABLE ACIDS. VOLATILE
ORGANIC ACIDS. HUMUS.

Carbohydrates, or Bodies of the Cellulose Group.—The steps by which organic matters become incorporated with the soil have been recounted on p. 135. When plants perish, their proximate principles become mixed with the soil. These organic matters shortly begin to decay or to pass into humus. In most circumstances, however, the soil must contain, temporarily or periodically, unaltered carbohydrates. Cellulose, especially, may be often found in an unaltered state in the form of fragments of straw, etc.

De Saussure (*Recherches*, p. 174) found that water dissolved from a rich garden soil that had been highly manured for a long time, several thousandths of organic matter, giving an extract, which, when concentrated, had an almost syrupy consistence and a sweet taste, was neither acid nor alkaline in reaction, and comported itself not unlike an impure mixture of glucose and dextrin.

Verdeil and Risler have made similar observations on ten soils from the farm of the *Institut Agronomique*, at Versailles. They found that the water-extract of these soils contained, on the average, 50%, of organic matter, which, when strongly heated, gave an odor like burning paper or sugar. These observers make no mention of crenates or apocrenates, and it, perhaps, remains somewhat doubtful, therefore, whether their researches really demonstrate the presence in the soil of a neutral body identical with, or allied to, dextrin or sugar.

Cellulose, starch, and dextrin, pass by fermentation into sugar (glucose); this may be resolved into lactic acid (the acid of sour milk and sour-krout), butyric acid (one of the acids of rancid butter), and acetic acid (the acid of vinegar). It must often happen that the bodies of the cellulose group ferment in the soil, the same as in the souring of milk or of dough, though they suffer for the most part conversion into humus, as will be shortly noticed.

Vegetable Acids, viz., oxalic, malic, tartaric, and citric acids, become ingredients of the soil when vegetable matters are buried in it. When the leaves of beets, tobacco, and other large-leaved plants fall upon the soil, oxalic and malic acids may pass into it in considerable quantity. Falling fruits may give it citric, malic, and tartaric acids. These acids, however, speedily suffer chemical change when in contact with decaying albuminoids. Buchner has shown (*Ann. Ch. u. Ph.*, 78, 207) that the solutions of salts of the above-named vegetable acids are rapidly converted into carbonates when mixed with vegetable ferments. In this process tartaric and citric acids are first partially converted into acetic acid, and this subsequently passes into carbonic acid.

Volatile Organic Acids.—Formic, propionic, acetic, and butyric acids, or rather their salts, have been detected by Jongbloed and others in garden earth. They are common products of fermentation, a process that goes on in the juices of plants that have become a part of the soil or of a compost.

These acids can scarcely exist in the soil, except temporarily, as results of fermentation or decay, and then in but very minute quantity. They consist of carbon, hydrogen, and oxygen. Their salts are all freely soluble in water. Their relations to agricultural plants have not been studied.

Humus (in part).—The general nature and origin of humus has been already considered. It is the *débris* of vegetation (or of animal matters) in certain stages of decomposition. Humus is considerably complex in its chemical character, and our knowledge of it is confessedly incomplete. In the paragraphs that immediately follow, we shall give from the best sources an account of its non-nitrogenous ingredients, so far they are understood, reserving to a later chapter an account of its nitrogenized constituents.

The Non-nitrogenous Components of Humus.—The appearance and composition of humus is different, according to the circumstances of its formation. It has already been mentioned that humus is brown or black in color. It appears that the first stage of decomposition yields the brown humus. It is seen in the dead leaves hanging to a tree in autumn, in the upper layers of fallen leaves, in the outer bark of trees, in the smut of wheat, and in the upper, dryer portions of peat.

When brown humus remains wet and with imperfect access of air, it decomposes further, and in time is converted into black humus. Black humus is invariably found in the soil beyond a little depth especially if it be compact, in the deeper layers of peat, in the interior of compost heaps, in the lower portions of the leaf-mould of forests, and in the mud or muck of swamps and ponds.

Ulmic Acid and Ulmin.—The brown humus contains

(besides, perhaps, unaltered vegetable matters) two characteristic ingredients, which have been designated *ulmic acid* and *ulmin*, (so named from having been found in a brown mass that exuded from an elm tree, *ulmus* being the Latin for elm). These two bodies demand particular notice.

When brown peat is boiled with water, it gives a yellowish or pale-brown liquid, being but little soluble in pure water. If, however, it be boiled with dilute solution of carbonate of soda (sal-soda), a dark-brown liquid is obtained, which owes its color to ulmate of soda. The alkali dissolves the insoluble ulmic acid by combining with it to form a soluble compound. By repeatedly heating the same portion of peat with new quantities of sal-soda solution, and pouring off the liquids each time, there arrives a moment when the peat no longer yields any color to the solution. The brown peat is thus separated into one portion soluble, and another insoluble, in carbonate of soda. *Ulmic acid* has passed into the solution, and *ulmin** remains undissolved (mixed, it may be, with unaltered vegetable matters, recognizable by their form and structure, and with sand and mineral substances).

By adding hydrochloric acid to the brown solution as long as it foams or effervesces, the ulmic acid separates in brown, bulky flocks, and is insoluble in dilute hydrochloric acid, but is a little soluble in pure water. When moist, it has an acid reaction, and dissolves readily in alkalies or alkali-carbonates. On drying, the ulmic acid shrinks greatly and remains as a brown, coherent mass.

The *ulmin** which remains after treatment of brown peat with carbonate of soda is an indifferent, neutral (i. e., not acid) body, which has the same composition as the

---

* The above statement is made on the authority of Mulder. The writer has, however, found, in several cases, that continued treatment with carbonate of soda alone completely dissolves the humus, leaving a residue of cellulose which yields nothing to caustic alkali. He is, therefore, inclined to disbelieve in the existence of ulmin and humin as distinct from ulmic and humic acids.

ulmic acid. By boiling it with caustic soda or potash-lye, it is converted without change of composition into *ulmic acid*.

On gently heating sugar with dilute hydrochloric acid, a brown substance is produced, which appears to be identical with the ulmic acid obtained from peat.

Humic Acid and Humin. — By treating *black humus* with carbonate of soda as above described, it is separated into *humic acid* and *humin**, which closely resemble ulmic acid and ulmin in all their properties—possess, however, a black color, and, as it appears, a somewhat different composition.

Humic acid and humin may be obtained also by the action of hot and strong hydrochloric acid, of sulphuric acid, and of alkalies, upon sugar and the other members of the cellulose group.

Composition of Ulmin, Ulmic Acid, Humin, and Humic Acid.—The results of the analyses of these bodies, as obtained by different experimenters and from different sources, are not in all cases accordant. Either several distinct substances have been confounded under each of the above names, or the true ulmin and humin, and ulmic and humic acids, are liable to occur mixed with other matters, from which they cannot be or have not been perfectly separated.

Mulder (*Chemie der Ackerkrume*, 1, p. 322), who has chiefly investigated these substances, believes there is a group of bodies having in general the characters of ulmin and ulmic acid, whose composition differs only by the elements of water,† and is exhibited by the general formula

$$C_{40} H_{28} O_{12} + nH_2O,$$

in which nH_2O signifies one, two, three, or more of water

---

* See note on page 225.
† In a way analogous to what is known of the sugars. (H. C. G., p. 80.)

Ulmic acid from sugar has the following composition in 100 parts:

Carbon,	67.1
Hydrogen,	4.2
Oxygen,	28.7
	100.0

which corresponds to $C_{40} H_{28} O_{12} H_2O$.

Mulder considers that in the same manner there exist various kinds of humic acids and humin, differing from each other by the elements of water, all of which may be represented by the general formula $C_{40} H_{24} O_{12} nH_2O$.

Humic acid and humin from sugar, corresponding to $C_{40} H_{24} O_{12} + 3H_2O$, have, according to Mulder, the following composition per cent:

Carbon,	64
Hydrogen,	4
Oxygen,	32
	100

Apocrenic and Crenic Acids.—In the acid liquid from which ulmic or humic acid has been separated, exist two other acids which were first discovered by Berzelius in the Porla spring in Sweden, and which bear the names *apocrenic acid* and *crenic acid* respectively. By adding soda to the acid liquid until the hydrochloric acid is neutralized, then acetic acid in slight excess, and lastly solution of acetate of copper (crystallized verdigris) as long as a dirty-gray precipitate is formed, the apocrenic acid is procured in combination with copper and ammonia. From this salt the acid itself may be separated* as a brown, gummy mass, which is easily soluble in water. According to Mulder it has the formula $C_{24} H_{12} O_{12} + H_2O$, or, in 100 parts,

---

* By precipitating the copper with sulphuretted hydrogen.

Carbon,   56.47
Hydrogen,   2.75
Oxygen,   40.78
———
100.00

Crenate of copper is lastly precipitated as a grass-green substance by adding acetate of copper to the liquid from which the apocrenate of copper was separated, and then neutralizing the free acid with ammonia. From this compound crenic acid may be prepared as a white, solid body of sour taste, to which Mulder ascribes the formula $C_{24} H_{24} O_{16} + 3H_2O$, and in 100 parts the following composition:

Carbon,   45.70
Hydrogen,   4.80
Oxygen,   49.50
———
100.00

Mutual Conversion of Apocrenic and Crenic Acids. —When, on the one hand, apocrenic acid is placed in contact with zinc and dilute sulphuric acid, the hydrogen evolved from the latter converts the brown apocrenic acid (by uniting with a portion of its oxygen) into colorless crenic acid. On the other hand, the solution of crenic acid exposed to the air shortly becomes brown by absorption of oxygen and formation of apocrenic acid. These changes may be repeated many times with the same portion of these substances.

Mulder remarks (*Chemie der Ackerkrume*, p. 350): "In every fertile soil these acids always occur together in not inconsiderable quantities. When the earth is turned over by the plow, two essentially different processes follow each other: oxidation, where the air has free access; reduction, where its access is more or less limited by the adhesion of the particles and especially by moisture. In the loose, dry earth apocrenic acid is formed; in the firm, moist soil, and in every soil after rain, crenic acid is produced, so that the action or effects of these substances are alternately manifested."

The Humus Bodies Artificially Produced. — When sugar, cellulose, starch, or gum, is boiled with strong hydrochloric acid or a strong solution of potash, brown or black bodies result which have the greatest similarity with the ulmin and humin, the ulmic and humic acids of peat and of soils.

By heating humus with nitric acid (a vigorous oxidizing agent), crenic and apocrenic acids are formed. The production of these bodies by such artificial means gives interesting confirmation of the reality of their existence, and demonstrates the correctness of the views which have been advanced as to their origin.

While the precise composition of all these substances may well be a matter of doubt, and from the difficulties of obtaining them in the pure state is likely to remain so, their existence in the soil and their importance in agricultural science are beyond question, as we shall shortly have opportunity to understand.

The Condition of these Humus Bodies in the Soil requires some comment. The organic substances thus noticed as existing in the soil are for the most part acids, but they do not exist to much extent in the free state, except in bogs and morasses. A soil that is fit for agricultural purposes contains little or no free acid, except carbonic acid, and oftentimes gives an alkaline reaction with test-papers.

Regarding ulmic and humic acids, which, as we have stated, are extracted by solution of carbonate of soda from humus, it appears that they do not exhibit acid characters before treatment with the alkali. They appear to be *altered* by the alkali and converted through its influence into acids. Only those portions of these bodies which are acted upon by the carbonates of potash, soda, and lime, that become ingredients of the soil by the solution of rocks, or by carbonate of ammonia brought down from the atmosphere or produced by decay of nitrogenous matters, acquire solubility, and are, in fact, acids; and these portions are acids in combination (salts), and not in the free state.

The Salts of the Humus Acids that may exist in the soil, viz., the ulmates, humates, apocrenates, and crenates of potash, soda, ammonia, lime, magnesia, iron, manganese, and alumina, require notice.

The *ulmates* and *humates* agree closely in their characters so far as is known.

The ulmates and humates of the alkalies (potash, soda, and ammonia) are *freely soluble in water*. They are formed when the alkalies or their carbonates come in contact 1st, with the ulmic and humic acids themselves; 2d, with the ulmates and humates of lime, magnesia, iron, and manganese; and 3d, by the action of the alkalies and their carbonates on humin and ulmin. Their solutions are yellow or brown.

The ulmates and humates of lime, magnesia, iron, manganese, and alumina, are *insoluble, or but very slightly soluble in water.*

From ordinary soils where these earths and oxides predominate, water removes but traces of humates and ulmates.

From peat, garden earth, and leaf-mould, which contain excess of the humic and ulmic acids, and carbonate of ammonia resulting from the decay of nitrogenous matters, water extracts a perceptible amount of these acids rendered soluble by the alkali.

There appear to exist *double salts* of humic acid and of ulmic acid, i. e., salts containing the acid combined with two or more bases. By adding solutions of compounds (e. g., sulphates) of lime, magnesia, iron, manganese, and alumina to solutions of humates or ulmates of the alkalies, precipitates are formed in which the acid is combined both with an alkali and an earth or oxide. These double salts are *insoluble or nearly so in water.*

Solutions of alkalies and alkali carbonates decompose them into soluble alkali humates or ulmates, and the earths or oxides are at least partially held in solution by the resulting compounds.

Mulder describes the following experiments, which justify the above conclusions. "Garden-soil was extracted with dilute solution of carbonate of soda, the soil being in excess. The solution was filtered and precipitated by addition of water, and the precipitate was washed and dissolved in a little ammonia. Thus was obtained a dark-brown solution of neutral humate of ammonia. The solution was rendered perfectly colorless by addition of caustic lime—basic humate of lime is therefore perfectly insoluble in water.

"Chloride of calcium rendered the solution very nearly colorless—neutral humate of lime is almost entirely insoluble.

"Calcined magnesia decolorized the solution perfectly. Chloride of magnesium made the solution very nearly colorless.

"The sulphates of protoxide and peroxide of iron, and sulphate of manganese, decolorized the solution perfectly.

"These decolorized liquids were made brown again by agitating them and the precipitated humates with carbonate of ammonia."

Apocrenates and Crenates.—According to Mulder, the crenates and apocrenates of the soil nearly always contain ammonia—are, in fact, double salts of this alkali with lime, iron, etc.

The *apocrenates* of the alkalies are freely soluble; those of the oxides of iron and manganese are moderately soluble; those of lime, magnesia, and alumina, are insoluble.

The *crenates* of the alkalies, of lime, magnesia, and protoxide of iron, are soluble; those of protoxide of iron and manganese are less soluble; crenate of alumina is insoluble.

All the salts of these acids that are insoluble of themselves are decomposed by, and soluble in, excess of the alkali-salts.

Do the Organic Matters of the Soil Directly Nourish Vegetation?—This is a question which, so far as humus is concerned, has been discussed with great earnestness by the most prominent writers on Agricultural Science.

De Saussure, Berzelius, and Mulder, have argued in the affirmative; while Liebig and his numerous adherents totally deny to humus the possession of any nutritive value. It is probable that humus may be directly absorbed by, and feed, plants. It is certain, also, that it does not contribute largely to the sustenance of agricultural crops.

To ascertain the real extent to which humus is taken up by plants, or even to demonstrate that it is taken up by them, is, perhaps, impossible from the data now in our possession. We shall consider the probabilities.

There have not been wanting attempts to ascertain experimentally whether humus is capable of feeding vegetation. Hartig, De Saussure, Wiegmann and Polstorf, and Soubeiran, have observed the growth of plants whose roots were immersed in solutions of humus. The experiments of Hartig led this observer to conclude that humate of potash and water-extract of peat do not enter the roots of plants. Not having had access to the original account of this investigation, the writer cannot, perhaps, judge properly of its merits. It appears, however, that the roots of the plants operated with were not kept constantly moist, and their extremities were decomposed by too great concentration of the liquid in which they were immersed. Under such conditions accurate results were out of the question.

De Saussure (*Ann. Ch. u. Ph.*, 42, 275) made two experiments, one with a bean, the other with *Polygonum Persicaria*, in which these plants were made to vegetate with their roots immersed in a solution of *humate of potash* (prepared by boiling humus with bicarbonate of potash). In the first case the bean plant, originally weighing 11 grams, gained during 14 days 6 grms., while the weight of the humus decreased 9 milligrams. The *Polygonum* during 10 days gained 3,5 grms., and the solution lost 43 milligrams of humus. These experiments Liebig considers undecisive, because an alkali-humate loses weight by oxidation (to carbonic acid and water) when exposed in solution to the air. Mulder, however, denies that any appreciable loss could occur in such a solution during the time of experiment, and considers the trials conclusive.

In a third experiment, De Saussure placed the roots of *Polygonum Persicaria* in the *water-extract* of turf containing no humic acid but crenic and apocrenic acids, where they remained nine days in a very flourishing state, putting forth new roots of a healthy white color. An equal quantity of the same extract was placed in a similar vessel for purposes of comparison. It was found that the solution in which the plants were stationed became paler in color and remained perfectly clear, while the other solution retained its original dark tint and became turbid. The former left after evaporation 33 mgrms., the latter 39 mgrms. of solid residue. The difference of 6 mgrms., De Saussure believes to have been absorbed by the plant.

Wiegmann and Polstorf (*Ueber die unorganischen Bestandtheile der Pflanzen*) experimented in a similar manner with *Mentha undulata*, a kind of mint, and *Polygonum Persicaria*, using two plants of 8 inches height, whose roots were well developed and perfectly healthy. The plants grew for 30 days in a wine-yellow *water-extract* of leaf compost (containing 148 mgrms. of solid substance—organic matter, carbonate of lime, etc.,—in 100 grams of extract), the roots being shielded from light, and during the same time an equal quantity of the same solution stood near by in a vessel of the same dimensions. The plants grew well, increasing $6\frac{1}{2}$ inches in length, and put forth long roots of a healthy white color. On the 18th of July the plants were removed from the solution, and 100 grams of the solution left on evaporation 132 mgrms. of residue. The same amount of humus extract, that had been kept in a contiguous vessel containing no plant, left a residue of 136 mgrms. The disappearance of humus from the solution is thus mostly accounted for by its oxidation.

De Saussure considers that his experiments demonstrate that humic acid and (in his third exp.) the matters extracted from peat by water (crenic and apocrenic acids) are absorbed by plants. Wiegmann and Polstorf attribute any apparent absorption in their trials to the unavoidable errors of experiment. The quantities that may have been absorbed were indeed small, but in our judgment not smaller than ought to be estimated with certainty.

Other experiments by Soubeiran, Malaguti, and Mulder, are on record, mostly agreeing in this, viz., that agricultural plants (beans, oats, cresses, peas, barley) grow well when their roots are immersed in, or watered by, solutions of humates, ulmates, crenates, and apocrenates of ammonia and potash. These experiments are, however, all unadapted to *demonstrate* that humus is absorbed by plants, and the trials of De Saussure and of Wiegmann, and Polstorf, are the only ones that have been made under conditions at all satisfactory to a just criticism. These do not, perhaps, conclusively demonstrate the nutritive function of humus. It is to be observed, however, that what evidence they do furnish is in its favor. They prove effectually that humus is not injurious to plants, though Liebig and Wolff have strenuously insisted that it is poisonous.

Let us now turn to the probabilities bearing on the question.

In the first place there are plants—those living in bogs and flourishing in dung-heap liquor—which throughout the whole period of their growth must *tolerate*, if not absorb, somewhat strong solutions of humus.

Again, the cultivated soil invariably yields some humus

(we use this word as a general collective term) to rain-water, and the richer the soil, as made so by manures and judged of by its productiveness, the larger the quantity, up to certain limits, of humus it contains. If, as we have seen, plants always contain silica, though this element be not essential to their development (II. C. G., p. 186), is it probable that they are able to reject humus so constantly presented to them under such a variety of forms?

Liebig opposes the view that humus contributes directly to the nourishment of plants because it and its compounds are insoluble; in the same book, however, (*Die Chemie in ihrer Anwendung auf Agricultur und Physiologie*, 7th Ed., 1862) he teaches the doctrine that all the food of the agricultural plant exists in the soil in an insoluble form. This old objection, still maintained, tallies poorly with his new doctrine. The old objection, furthermore, is baseless, for the humates are as soluble as phosphates, which are gathered by every plant and from all soils.

It has been the habit of Liebig and his adherents to teach that the plant is nourished exclusively by the last products of the destruction of organic matter, viz., by carbonic acid, ammonia, nitric acid, and water, together with the ingredients of ashes. While no one denies or doubts that these substances chiefly nourish agricultural plants, no one can deny that other bodies may and do take part in the process. It is well established that various organic substances of animal origin, viz., urea, uric acid, and glycocoll, are absorbed by, and nourish, agricultural plants; while it is universally known that the principal food of multitudes of the lower orders of plants, the fungi, including yeast, mould, rust, brand, mushrooms, are fed entirely, so far as regards their carbon, on organic matters. Thus, yeast lives upon sugar, the vinegar plant on acetic acid, the *Peronospora infestans* on the juices of the potato, etc. There are many parasitic plants of a higher order common in our forests whose roots are fastened upon and absorb the juices of the roots of trees; such are the beech drops (*Epiphegus*), pine drops (*Pterospora*), Indian pipe (*Monotropa*); the last-named also grows upon decayed vegetable matter.

The dodder (*Cuscuta*) is parasitic upon living plants, especially upon flax, whose juices it appropriates often to the destruction of the crop.

It is indeed true that there is a wide distinction between most of these parasites and agricultural plants. The former are mostly destitute of chlorophyll, and appear to be totally incapable of assimilating carbon from carbonic acid.* The latter acquire certainly the most of their food from carbonic acid, but in their root-organs they contain no chlorophyll; there they cannot assimilate carbon from carbonic acid. They do assimilate nitrogen from the organic principles of urine; what is to hinder their obtaining carbon from the soluble portions of humus, from the organic acids, or even from unaltered carbohydrates?

De Saussure, in his investigation just quoted from, says further: "After having thus demonstrated † the absorption of humus by the roots, it remains to speak of its assimilation by the plant. One of the indications of this assimilation is derived from the absence of the peculiar color of humus in the interior of the plant, which has absorbed a strongly colored solution of humate of potash, as compared to the different deportment of coloring matters

---

* Dr. Luck (*Ann. Chem. u. Pharm.*, 78, 85) has indeed shown that the mistletoe (*Viscum album*) decomposes carbonic acid in the sunlight, but this plant has greenish-yellow leaves containing chlorophyll.

† We take occasion here to say explicitly that the only valid criticism of De Saussure's experiment on the *Polygonum* supplied with humate of potash, is Liebig's, to the effect that the solution lost humic acid to the amount of 43 milligrams not as a result of absorption by the plant, but by direct oxidation. Mulder and Soubeiran both agree that such a solution could not lose perceptibly in this way. That De Saussure was satisfied that such a loss could not occur, would appear from the fact that he did not attempt to estimate it, as he did in the subsequent experiment with water-extract of peat. If, now, Liebig be wrong in his objection (and he has furnished no proof that his statement is true), then De Saussure *has demonstrated* that humic acid is absorbed by plants.

(such as ink) which cannot nourish the plant. The latter (ink, etc.) leave evidences of their entrance into the plant, while the former are changed and partly assimilated.

"A bean 15 inches high, whose roots were placed in a decoction of Brazil-wood (to which a little alum had been added and which was filtered), was able to absorb no more than the fifth part of its weight of this solution without wilting and dying. In this process four-fifths of its stem was colored red.

"*Polygonum Persicaria* (on occasion an aquatic or bog plant) grew very well in the same solution and absorbed its coloring matter, but the color never reached the stem. The red principle of Brazil-wood being partially assimilated by the *Polygonum*, underwent a chemical change; while in the bean, which it was unable to nourish, it suffered no change. The *Polygonum* itself became colored, and withered when its roots were immersed in diluted ink."

Biot (*Comptes Rendus*, 1837, 1, 12) observed that the red juice of *Phytolacca decandra* (poke-weed), when poured upon the soil in which a white hyacinth was blossoming, was absorbed by the plant, and in one to two hours dyed the flowers of its own color. After two or three days, however, the red color disappeared, the flowers becoming white again.

From the facts just detailed, we conclude that some kinds of organic matters may be absorbed and chemically changed (certain of them assimilated) by agricultural plants.

We must therefore hold it to be extremely probable that various forms of humus, viz., soluble humates, ulmates, crenates, and apocrenates, together with the other soluble organic matters of the soil, are taken up by plants, and decomposed or transformed, nay, we may say, assimilated by them.

A few experiments might easily be devised which would completely settle this point beyond all controversy.

Organic Matters as Indirect Sources of Carbon to Plants.—The decay of organic matters in the soil supplies to vegetation considerably more carbonic acid *in a given time* than would be otherwise at the command of crops. The quantities of carbonic acid found in various soils have already been given (p. 219). The beneficial effects of such a source of carbonic acid in the soil are sufficiently obvious (p. 128).

Organic Matters not Essential to the Growth of Crops.—Although, on the farm, crops are rarely raised without the concurrence of humus or at least without its presence in the soil, it is by no means indispensable to their life or full development. Carbonic acid gas is of itself able to supply the rankest vegetation with carbon, as has been demonstrated by numerous experiments, in which all other compounds of this element have been excluded (p. 48).

§ 4.

THE AMMONIA OF THE SOIL.

In the chapter on the Atmosphere as the food of plants we have been led to conclude that the element *nitrogen*, so indispensable to vegetation as an ingredient of albumin, etc., is supplied to p'ants *exclusively by its compounds, and mainly by ammonia and nitric acid,* or by substances which yield these bodies readily on oxidation or decay.

We have seen further that both ammonia and nitric acid exist in very minute quantities in the atmosphere, are dissolved in the atmospheric waters, and by them brought into the soil.

It is pretty fairly demonstrated, too, that these bodies, as occurring in the atmosphere, become of appreciable use to agricultural vegetation only after their incorporation with the soil.

Rain and dew are means of collecting them from the atmosphere, and, as we shall shortly see, the soil is a storehouse for them and the medium of their entrance into vegetation.

This is therefore the proper place to consider in detail the origin and formation of ammonia and nitric acid, so far as these points have not been noticed when discussing their relations to the atmosphere.

Ammonia is formed in the Soil either in the decay of organic bodies containing nitrogen, as the albuminoids, etc., or by the reduction of nitrates (p. 74). The former process is of universal occurrence since both vegetable and animal remains are constantly present in the soil; the latter transformation goes on only under certain conditions, which will be considered in the next section (p. 269).

The statement that ammonia is generated from the free nitrogen of the air and the nascent hydrogen of decomposing carbohydrates, as cellulose, starch, etc., or that set free from water in the oxidation of certain metals, as iron and zinc, has been completely disproved by Will. (*Ann. d. Ch. u. Ph.*, 45, pp. 106-112.)

The ammonia encountered in such experiments may have been, 1st, that pre-existing in the pores of the substances, or dissolved in the water operated with. Faraday (*Researches in Chemistry and Physics*, p. 143) has shown by a series of exact experiments that numerous, we may say all, porous bodies exposed to the air have a minute amount of ammonia adhering to them; 2d, that which is generated in the process of testing or experimenting (as when iron is heated with potash), and formed by the action of an alkali on some compound of nitrogen occurring in the materials of the experiment; or, 3d. that which results from the reduction of a nitrite formed from free nitrogen by the action of ozone (pp. 77-83).

The Ammonia of the Soil.—*a. Gaseous Ammonia as Carbonate.*—Boussingault and Lewy, in their examination of the air contained in the interstices of the soil, p. 219, tested it for ammonia. In but two instances did they find sufficient to weigh. In all cases, however, they were able to detect it, though it was present in very minute quantity. The two experiments in which they were able to weigh the ammonia were made in a light, sandy soil from which potatoes had been lately harvested. On the 2d of September the field was manured with stable dung; on the 4th the first experiment was made, the air being taken, it must be inferred from the account given, at a depth of 14 inches. In a million parts of air by weight were found 82 parts of ammonia. Five days subsequently, after rainy weather, the air collected at the same place contained but 13 parts in a million.

b. Ammonia physically condensed in the Soil.—Many porous bodies condense a large quantity of ammonia gas. Charcoal, which has an extreme porosity, serves to illustrate this fact. De Saussure found that box-wood charcoal, freshly ignited, absorbed 98 times its volume of ammonia gas. Similar results have been obtained by Stenhouse, Angus Smith, and others (p. 166). The soil cannot, however, ordinarily contain more than a minute quantity of physically absorbed ammonia. The reasons are, first, a porous body saturated with ammonia loses the greater share of this substance when other gases come in contact with it. It is only possible to condense in charcoal 98 times its volume of ammonia, by cooling the hot charcoal in mercury which does not penetrate it, or in a vacuum, and then bringing it directly into the *pure* ammonia gas. The charcoal thus saturated with ammonia loses the latter rapidly on exposure to the air, and Stenhouse has found by actual trial that charcoal exposed to ammonia and afterwards to air retains but minute traces of the former. Secondly, the soil when adapted for vegetable growth is moist or wet. The water of the soil which covers the particles of earth, rather than the particles themselves, must contain any absorbed ammonia. Thirdly, there are in fertile soils substances which combine chemically with ammonia.

That the soil does contain a certain quantity of ammonia adhering to the surface of its particles, or, more probably, dissolved in the hygroscopic water, is demonstrated by the experiments of Boussingault and Lewy just alluded to, in all of which ammonia was detected in the air included in the cavities of the soil. In case ammonia were physically condensed or absorbed, a portion of it would be carried off in a current of air in the conditions of Boussingault and Lewy's experiments,—nay, all of it would be removed by such treatment sufficiently prolonged.

Brustlein (Boussingault's *Agronomie*, etc., 1, p. 152) records that 100 parts of moist earth placed in a vessel of about 2½ quarts capacity containing 0.9 parts of (free) ammonia, absorbed during 3 hours a little more than 0.4 parts of the latter. In another trial 100 parts of the same earth dried, placed under the same circumstances, absorbed 0.28 parts of ammonia and 2.6 parts of water.

Brustlein found that soil placed in a confined atmosphere containing very limited quantities of ammonia cannot condense the latter completely. In an experiment similar to those just described, 100 parts of earth (tenacious calcareous clay) and 0.019 parts of ammonia were left together 5 days. At the conclusion of this period 0.016 parts of the latter had been taken up by the earth. The remainder was found to be dissolved in the water that had evaporated from the soil, and that formed a dew on the interior of the glass vessel.

Brustlein proved further that while air may be almost entirely deprived of its ammonia by traversing a long column of soil, so the soil that has absorbed ammonia readily gives up a large share of it to a stream of pure air. He caused air, charged with ammonia gas by being made to bubble through water of ammonia, to traverse a tube 1 ft. long filled with small fragments of moist soil. The ammonia was completely absorbed in the first part of the experiment. After about 7 cubic feet of air had streamed through the soil, ammonia began to escape unabsorbed. The earth thus saturated contained 0.192°|₀ of ammonia. A current of pure air was now passed through the soil as long as ammonia was removed by it in notable quantity, about 38 cubic feet being required. By this means more than one-half the ammonia was displaced and carried off, the earth retaining but 0.084°|₀.

Brustlein ascertained further that ammonia which has been absorbed by a soil from aqueous solution escapes easily when the earth is exposed to the air, especially when it is repeatedly moistened and allowed to dry.

100 parts of the same kind of soil as was employed in the experiments already described were agitated with 187 parts of water containing 0.889 parts of ammonia. The earth absorbed 0.157 parts of ammonia. It was now drained from the liquid and allowed to dry at a low temperature, which operation required eight days. It was then moistened and allowed to dry again, and this was repeated four times. The progressive loss of ammonia is shown by the following figures.

```
100 parts of soil absorbed ........................0.157 parts of ammonia.
 "    "    "   "  contained after the first  drying........0.083  "    "    "
 "    "    "   "       "        "    "  second   "      ........0.066  "    "    "
 "    "    "   "       "        "    "  third    "      ........0.051  "    "    "
 "    "    "   "       "        "    "  fourth   "      ........0.011  "    "    "
 "    "    "   "       "        "    "  fifth    "      ........0.039  "    "    "
```

In this instance the loss of ammonia amounted to three-fourths the quantity at first absorbed.

The extent to which absorbed ammonia escapes from the soil is greatly increased by the evaporation of water. Brustlein found that a soil containing 0.067°|₀ of ammonia suffered only a trifling loss by keeping 43 days in a dry place, whereas the same earth lost half its ammonia in a shorter time by being thrice moistened and dried.

According to Knop (*Vs. St.*, III, p. 222), the single

proximate ingredient of soils that under ordinary circumstances exerts a considerable surface attraction for ammonia gas is *clay*. Knop examined the deportment of ammonia in this respect towards sand, soluble silica, pure alumina, carbonate of lime, carbonate of magnesia, hydrated sesquioxide of iron, sulphate of lime, and humus.

To recapitulate, the soil contains carbonate of ammonia physically absorbed in its pores, i. e., adhering to the surfaces of its particles,—as Knop believes, to the particles of *clay*. The quantity of ammonia is variable and constantly varying, being increased by rain and dew, or manuring, and diminished by evaporation of water. The actual quantity of physically absorbed ammonia is, in general, very small, and an accurate estimation of it is, perhaps, impracticable, save in a few exceptional cases.

c. Chemically combined Ammonia.—The reader will have noticed that in the experiments of Brustlein just quoted, a greater quantity of ammonia was absorbed by the soil than afterwards escaped, either when the soil was subjected to a current of air or allowed to dry after moistening with water. This ammonia, it is therefore to be believed, was in great part retained in the soil in chemical combination in the form of compounds that not only do not permit it readily to escape as gas, but also are not easily washed out by water. The bodies that may unite with ammonia to comparatively insoluble compounds are, 1st, the organic acids of the humus group*—the humus acids, as we may designate them collectively. The salts of these acids have been already noticed. Their com-

---

* Mulder asserts that the affinity of ulmic, humic, and apocrenic acids for ammonia is so strong that they can only be freed from it by evaporation of their solutions to dryness with caustic potash. Boiling with carbonate of potash or carbonate of soda will not suffice to decompose their ammonia-salts. We hold it more likely that the ammonia which requires an alkali for its expulsion is generated by the decomposition of the organic acid itself, or, if that be destitute of nitrogen, of some nitrogenous substance admixed. According to Boussingault, ammonia is completely removed from humus by boiling with water and caustic magnesia.

pounds with ammonia are freely soluble in water; hence strong solution of ammonia dissolves them from the soil. But when ammonia salts of these acids are put in contact with lime, magnesia, oxide of iron, oxide of manganese, and alumina, the latter being in preponderating quantity, there are formed double compounds which are insoluble or slightly soluble. Since the humic, ulmic, crenic, and apocrenic acids always exist in soils which contain organic remains, there can be no question that these double salts are a chemical cause of the retention of ammonia in the soil.

2d. Certain phosphates and silicates hereafter to be noticed have the power of forming difficultly soluble compounds with ammonia.

Reserving for a subsequent chapter a further discussion of the causes of the chemical retention of ammonia in the soil, we may now appropriately recount the observations that have been made regarding the condition of the ammonia of the soil as regards its volatility, solubility, etc.

Volatility of the Ammonia of the Soil.—We have seen that ammonia may escape from the soil as gaseous carbonate. The fact is not only true of this substance as physically absorbed, but also under certain conditions of that chemically combined. When we mingle together equal bulks of sulphate of lime (gypsum) and carbonate of ammonia, both in the state of fine powder, the mixture begins and continues to smell strongly of ammonia, owing to the volatility of the carbonate. If now the mixture be drenched with water, the odor of ammonia at once ceases to be perceptible, and if, after some time, the mixture be thrown on a filter and washed with water, we shall find that what remains undissolved contains a large proportion of carbonate of lime, as may be shown by its dissolving in an acid with effervescence; while the liquid that has passed the filter contains sulphate of ammonia, as may be learned by the appropriate chemical tests or by evaporating to dryness, when it will remain as a colorless, odorless, crystalline solid. Double decomposition has taken place between the two salts under the influence of water. If, again, the carbonate of lime on the filter be reunited to the liquid filtrate and the whole be evaporated, it will be found that when the water has so far passed off that a moist, pasty mass remains, the odor of ammonia becomes evident again—carbonate of ammonia, in fact, escaping by volatilization, while sulphate of lime is reproduced. It is a general law in chemistry that when a number of acids and bases are together, those which under the circumstances can produce by their union a volatile body will unite, and those which under the circumstances can form a solid body will unite. When carbonic and sulphuric acids, lime and ammonia, are in mixture, it is the *circumstances* which determine in what mode these bodies combine. In presence of much water carbonate of lime is formed because of its insolubility, water not being able to destroy its solidity, and sulphate of ammonia necessarily results by the union of the other two substances. When the water is removed by evaporation, all the possible compounds between carbonic and sulphuric acids, lime and ammonia, become solid; the compound of ammonia and carbonic acid being then volatile, this fact determines its formation, and, as it escapes, the lime and sulphuric acid can but remain in combination.

To apply these principles: When carbonate of ammonia is brought into the soil by rain, or otherwise, it tends in presence of much water to enter into insoluble combinations so far as is possible. When the soil becomes dry, these compounds begin to undergo decomposition, provided carbonates of lime, magnesia, potash, and soda, are present to transpose with them; these bases taking the place of the ammonia, while the carbonic acid they were united with, forms with the latter a volatile compound. In this way, then, all soils, for it is probable that no soil exists which is destitute of carbonates, may give off at the surface in dry weather a portion of the ammonia which before was chemically retained within it.

Solubility of the Ammonia of the Soil.—The distinctions between physically adhering and chemically combined ammonia are difficult, if not impossible, to draw with accuracy. In what follows, therefore, we shall not attempt to consider them separately.

When ammonia, carbonate of ammonia, or any of the following ammoniacal salts, viz., chloride, sulphate, nitrate, and phosphate, are dissolved in water, and the solutions are filtered through or agitated with a soil, we find that a portion of ammonia is invariably removed from solution and absorbed by the soil. An instance of this absorbent action has been already given in recounting Brustlein's experiments, and further examples will be hereafter adduced when we come to speak of the silicates of the soil. The points to which we now should direct attention are these, viz., 1st, *the soil cannot absorb ammonia completely from its solutions ;* and, 2d, *the ammonia which it does absorb may be to a great degree dissolved out again by water.* In other words, the compounds of ammonia that are formed in the soil, though comparatively insoluble, are not absolutely so.

Henneberg and Stohmann found that a light, calcareous, sandy garden soil, when placed in twice its weight of pure water for 24 hours, yielded to the latter $\frac{1}{5000}$ of its weight of ammonia (=0.0002"|$_0$).

100 parts of the same soil left for 24 hours in 200 parts of a solution of chloride of ammonium (containing 2.182 of sal-ammoniac =0.693 part of ammonia), absorbed 0.112 part of ammonia. Half of the liquid was poured off and its place supplied with pure water, and the whole left for 24 hours, when half of this liquid was taken, and the process of dilution was thus repeated to the fifth time. In the portions of water each time removed, ammonia was estimated, and the result was that the water added dis-

olved out nearly one-half the ammonia which the earth at first absorbed.

The 1st dilution removed from the soil 0.010
" 2d   "   "   "   "   " 0.009
" 3d   "   "   "   "   " 0.014
" 4th   "   "   "   "   " 0.011
" 5th   "   "   "   "   " 0.009

Total 0.053

Deducting 0.053 from the quantity first absorbed, viz., 0.112, there remains 0.059 part retained by the soil after five dilutions. Knop, in 11 decantations, in which the soil was treated with 8 times its weight of water, removed $93°|_0$ of the ammonia which the soil had previously absorbed. We cannot doubt that by repeating the washing sufficiently long, all the ammonia would be dissolved, though a very large volume of water would certainly be needful.

Causes which ordinarily prevent the Accumulation of Ammonia in the Soil.—The ammonia of the soil is constantly in motion or suffering change, and does not accumulate to any great extent. In summer, the soil daily absorbs ammonia from the air, receives it by rains and dews, or acquires it by the decay of vegetable and animal matters.

Daily, too, ammonia wastes from the soil—by volatilization—accompanying the vapor of water which almost unceasingly escapes into the atmosphere.

When the soil is moist and the temperature not too low, its ammonia is also the subject of remarkable chemical transformations. Two distinct chemical changes are believed to affect it; one is its oxidation to nitric acid. This process we shall consider in detail in the next section. As a result of it, we never find ammonia in the water of ordinary wells or deep drains, but instead always encounter nitric acid united to lime, and, perhaps, to magnesia and alkalies. The other chemical change appears to be the alteration of the compounds of ammonia with the humus acids, whereby bodies result which are no longer soluble in water, and which, as such, are probably innutritious to plants. These substances are quite slowly decomposed when put in contact, especially when heated with alkalies or caustic lime in the presence of water. In this decomposition ammonia is reproduced. These indifferent nitrogenous matters appear to be analogous to a class of substances known to chemists as *amides*, of which asparagin, a crystallizable body obtained from asparagus, young peas, etc., and urea and uric acid, the characteristic ingredients of urine, are examples. Further account of these matters will be given subsequently, p. 276.

Quantity of Ammonia in Soils.—Formerly the amount of ammonia in soils was greatly overestimated, as the result of imperfect methods of analysis. In 1846, Krocker, at Liebig's instigation, estimated the nitrogen of 22 soils, and Liebig published some ingenious speculations in which all this nitrogen was incorrectly assumed to be in the form of ammonia. Later, various experimenters have attempted to estimate the ammonia of soils. In 1855, the writer examined several soils in Liebig's laboratory. The soils were boiled for some hours with water and caustic lime, or caustic potash. The ammonia that was set free, distilled off, and its amount was determined by alkalimetry. It was found that however long the distillation was kept up, ammonia continued to come over in minute quantity, and it was probable that this substance was not simply expelled from the soil, but was slowly formed by the action of lime on organic matters, it being well known to chemists that many nitrogenous bodies are thus decomposed. The results were as follows:

								Ammonia.
White sandy loam distilled with caustic lime gave in two Exp's.	0.0169 p.ct.
								0.0186 "
Yellow clay	"	"	"	"	"	"	"	0.0047 "
								0.0051 "
" "	"	"	"	potash	"	" one	"	0.0075 "
Black garden soil	"	"	"	lime	"	" two	"	0.0631 "
								0.0923 "

The fact that caustic potash, a more energetic decomposing agent than lime, disengaged more ammonia than the latter from the yellow clay, strengthens the view that ammonia is produced and not merely driven off under the conditions of these experiments, and that accordingly the figures are too high. Other chemists employing the same method have obtained similar results.

Boussingault (*Agronomie*, T. III, p. 206) was the first to substitute magnesia for potash and lime in the estimation of ammonia, having first demonstrated that this substance, so feebly alkaline, does not perceptibly decompose gelatine, albumin, or asparagine, all of which bodies, especially the latter, give ammonia when boiled with milk of lime or solutions of potash. The results of Boussingault here follow.

Localities.		Quantity of Ammonia per cent.
Liebfrauenberg,	Alsatia	0.0022
Bischwiller,	"	0.0020
Merckwiller,	"	0.0011
Bechelbronn,	"	0.0009
Mittelhausbergen,	"	0.0007
Ile Napoleon, Mulhouse,		0.0006
Argentan, Orne,		0.0060
Quesnoy-sur-Deule, Nord,		0.0012
Rio Madeira,	America,	0.0090
Rio Trombetto,	"	0.0030
Rio Negro,	"	0.0038
Santarem,	"	0.0083
Ile du Salut,	"	0.0080
Martinique,	"	0.0085
Rio Cupari, (leaf mold,)	"	0.0525
Peat,	Paris,	0.0180

The above results on French soils correspond with those obtained more recently on soils of Saxony by Knop and Wolff, who have devised an ingenious method of estimating ammonia, which is founded on altogether a different principle. Knop and Wolff measure the nitrogen gas which is set free by the action of chloride of soda (Javelle water*) in a specially constructed apparatus, the

---

* More properly *hypochlorite of soda*, which is used in mixture with bromine and caustic soda.

Azotometer. (*Chemisches Centralblatt*, 1860, pp. 243 and 534.)

By this method, which gives accurate results when applied to known quantities of ammonia-salts, Knop and Wolff obtained the following results:

$$\text{Ammonia in dry soil.}$$

Very light sandy soil from birch forest............0.00077°|$_0$
Rich lime soil from beech forest.................0.00087
Sandy loam, forest soil..........................0.00012
Forest soil.....................................0.00080
Meadow soil, red sandy loam.....................0.00027

Average.........................0.00056

The rich alluvial soils from tropical America are ten or more times richer in ready-formed ammonia than those of Saxony. These figures show then that the substance in question is very variable as a constituent of the soil, and that in the ordinary or poorer classes of unmanured soils its percentage is scarcely greater than in the atmospheric waters.

The Quantity of Ammonia fluctuates. — Boussingault has further demonstrated by analysis what we have insisted upon already in this chapter, viz., that the quantity of ammonia is liable to fluctuations. He estimated ammonia in garden soil on the 4th of March, 1860, and then, moistening two samples of the same soil with pure water, examined them at the termination of one and two months respectively. He found,

March 4th, 0.009°|$_0$ of ammonia.
April "   0.014 "   "   "
May   "   0.019 "   "   "

The simple standing of the moistened soil for two months sufficed in this case to double the content of ammonia.

The quantitative fluctuations of this constituent of the soil has been studied further both by Boussingault and by Knop and Wolff. The latter in seeking to answer the question—"How great is the ammonia-content of good manured soil lying fallow?"—made repeated determinations of ammonia (17 in all) in the same soil (well-manured, sandy, calcareous loam exposed to all rains and dews but not washed) during five months. The moist soil varied in its proportion of ammonia with the greatest irregularity between the extremes of 0.0008 and 0.0003°|₀. Similar observations were made the same summer on the loamy soil of a field, at first bare of vegetation, then covered with a vigorous potato crop. In this case the fluctuations ranged from 0.0009 to 0.0003°|₀ as irregularly as in the other instance.

Knop and Wolff examined the soil last mentioned at various depths. At 3 ft. the proportion of ammonia was scarcely less than at the surface. At 6 ft. this loam, and at a somewhat greater depth an underlying bed of sand, contained *no trace of ammonia.* This observation accords with the established fact that deep well and drainwaters are destitute of ammonia.

Boussingault has discovered (*Agronomie,* 3, 195) that the addition of caustic lime to the soil largely increases its content of ammonia—an effect due to the decomposing action of lime on the amide-like substances already noticed.

§ 5.

NITRIC ACID (NITRATES, NITROUS ACID, AND NITRITES) OF THE SOIL.

Nitric acid is formed in the atmosphere by the action of ozone, and is brought down to the soil occasionally in the free state, but almost invariably in combination with ammonia, by rain and dew, as has been already described (p. 86). It is also produced in the soil itself by processes whose nature—considerably obscure and little understood —will be discussed presently.

In the soil, nitric acid is always combined with an alkali or alkali-earth, and never exists in the free state in appreciable quantity. We speak of nitric acid instead of nitrates, because the former is the active ingredient common to all the latter. Before considering its formation and nutritive relations to vegetation, we shall describe those of its compounds which may exist in the soil, viz., the *nitrates of potash, soda, lime, magnesia, and iron.*

Nitrate of Potash ($K\ NO_3$) is the substance commercially known as niter or saltpeter. When pure (refined saltpeter), it occurs in colorless prismatic crystals. It is freely soluble in water, and has a peculiar sharp, cooling taste. Crude saltpeter contains common salt and other impurities. Nitrate of potash is largely procured for industrial uses from certain districts of India (Bengal) and from various caves in tropical and temperate climates, by simply leaching the earth with water and evaporating the solution thus obtained. It is also made in artificial niter-beds or plantations in many European countries. It is likewise prepared artificially from nitrate of soda and caustic potash, or chloride of potassium. The chief use of the commercial salt is in the manufacture of gunpowder and fireworks.

Sulphur, charcoal, (which are ingredients of gunpowder), and other combustible matters, when heated in contact with a nitrate, burn with great intensity at the expense of the oxygen which the nitrate contains in large proportion and readily parts with.

Nitrate of Soda ($Na\ NO_3$) occurs in immense quantities in the southern extremity of Peru, province of Tarapaca, as an incrustation or a compact stratum several feet thick, on the pampa of Tamarugel, an arid plain situated in a region where rain never falls. The salt is dissolved in hot water, the solution poured off from sand and evaporated to the crystallizing point. The crude salt has in general a yellow or reddish color. When pure, it is white or colorless. From the shape of the crystals it has been called cubic* niter; it is also known as Chili saltpeter, having been formerly exported from Chilian ports, and is sometimes termed soda-saltpeter. In 1854, about 40,000 tons were shipped from the port of Iquique.

Nitrate of soda is hygroscopic, and in damp air becomes quite moist, or even deliquesces, and hence is not suited for making gunpowder. It is easily procured artificially by dissolving carbonate of soda in nitric acid. This salt is largely employed as a fertilizer, and for preparing nitrate of potash and nitric acid.

Nitrate of Lime ($Ca2NO_3$) may be obtained as a white mass or as six-sided crystals by dissolving lime in nitric acid and evaporating the solution. It absorbs water from the air and runs to a liquid. Its taste is bitter and sharp. Nitrate of lime exists in well-waters and accompanies nitrate of potash in artificial niter-beds.

Nitrate of Magnesia ($Mg2NO_3$) closely resembles nitrate of lime in external characters and occurrence. It may be prepared by dissolving magnesia in nitric acid and evaporating the solution.

Nitrates of Iron.—Various compounds of nitric acid and iron, both soluble and insoluble, are known. In the soil it is probable that only insoluble basic nitrates of sesquioxide can occur. Knop observed (*V. St.*, V, 151) that certain soils when left in contact with solution of nitrate of potash for some time, failed to yield the latter entirely to water again. The soils that manifested this anomalous deportment were rich in humus, and at the same time contained much sesquioxide of iron that could be dissolved out by acids. It is possible that nitric acid entered into insoluble combinations here, though this hypothesis as yet awaits proof.

---

* The crystals are, in fact, rhomboidal.

Nitrates of alumina are known to the chemist, but have not been proved to exist in soils. Nitrate of ammonia has already been noticed, p. 71.

Nitric Acid not usually fixed by the Soil.—In its deportment towards the soil, nitric acid (either free or in its salts) differs in most cases from ammonia in one important particular. The nitrates are usually not fixed by the soil, but remain freely soluble in water, so that washing readily and completely removes them. The nitrates of ammonia and potash are decomposed in the soil, the alkali being retained, while the nitric acid may be removed by washing with water, mostly in the form of nitrate of lime. Nitrate of soda is partially decomposed in the same manner. Free nitric acid unites with lime, or at least is found in the washings of the soil in combination with that base.

As just remarked, Knop has observed that certain soils containing much organic matters and sesquioxide of iron, appeared to retain or decompose a small portion of nitric acid (put in contact with them in the form of nitrate of potash). Knop leaves it uncertain whether this result is simply the fault of the method of estimation, caused by the formation of basic nitrate of iron, which is insoluble in water, or, as is perhaps more probable, due to the decomposing (reducing) action of organic matters.

Nitrification is the formation of nitrates. When vegetable and animal matters containing nitrogen decay in the soil, nitrates of these bases presently appear. In Bengal, during the dry season, when for several months rain seldom or never falls, an incrustation of saline matters, chiefly nitrate of potash, accumulates on the surface of those soils, which are most fertile, and which, though cultivated in the wet season only, yield two and sometimes three crops of grain, etc., yearly. The formation of nitrates, which probably takes place during the entire year, appears to go on most rapidly in the hottest weather.

The nitrates accumulate near the surface when no rain falls to dissolve and wash them down—when evaporation causes a current of capillary water to ascend continually in the soil, carrying with it dissolved matters which must remain at the surface as the water escapes into the atmosphere. In regions where rain frequently falls, nitrates are largely formed in rich soils, but do not accumulate to any extent, unless in caves or positions artificially sheltered from the rain.

Boussingault's examination of garden earth from Liebfrauenberg (*Agronomie*, etc., T. II, p. 10) conveys an idea of the progress which nitrification may make in a soil under cultivation, and highly charged with nitrogenous manures. About 2.3 lbs. of sifted and well-mixed soil were placed in a heap on a slab of stone under a glazed roof. From time to time, as was needful, the earth was moistened with water exempt from ammonia. The proportion of nitric acid was determined in a sample of it on the day the experiment began, and the analysis was repeated four times at various intervals. The subjoined statement gives the per cent of nitrates expressed as nitrate of potash in the dry soil, and also the quantity of this salt contained in an acre taken to the depth of one foot.*

	Per cent.	Lbs. per acre.
1857— 5th August,	0.01	34
" —17th "	0.06	222
" — 2d September,	0.18	634
" —17th "	0.22	760
" — 2d October,	0.21	728

The formation of nitrates proceeded rapidly during the heat of summer, but ceased by the middle of September. Whether this cessation was due to the lower temperature or to the complete nitrification of all the matter existing in the soil capable of this change, or to decomposition of nitric acid by the reducing action of organic matters,

---

* The figures given above are abbreviated from the originals, or reduced to English denominations with a trifling loss of exactness.

further researches must decide. The quantities that accumulated in this experiment are seen to be very considerable, when we remember that experience has shown that 200 lbs. per acre of the nitrates of potash or soda is a large dressing upon grain or grass. Had the earth been exposed to occasional rain, its analysis would have indicated a much less percentage of nitrates, because the salt would have been washed down far into, and, perhaps, out of, the soil but no less, probably even somewhat more, would have been actually formed. In August, 1856, Boussingault examined earth from the same garden after 14 days of hot, dry weather. He found the nitrates equal to 911 lbs. of nitrate of potash per acre taken to the depth of one foot. From the 9th to the 29th of August it rained daily at Liebfrauenberg, more than two inches of water falling during this time. When the rain ceased, the soil contained but 38 lbs. per acre. In September, rain fell 15 times, and to the amount of four inches. On the 10th of October, after a fortnight of hot, windy weather, the garden had become so dry as to need watering. On being then analyzed, the soil was found to contain nitrates equivalent to no less than 1,290 lbs. of nitrate of potash per acre to the depth of one foot. This soil, be it remembered, was porous and sandy, and had been very heavily manured with well-rotted compost for several centuries.

Boussingault has examined more than sixty soils of every variety, and in every case but one found an appreciable quantity of nitrates. Knop has also estimated nitric acid in several soils (*Versuchs St.*, V, 143). Nitrates are almost invariably found in all well and river waters, and in quantities larger than exist in rain. We may hence assume that nitrification is a process universal to all soils, and that nitrates are normal, though, for the reasons stated, very variable ingredients of cultivated earth.

The Sources of the Nitric Acid which is formed within the Soil.—Nitric acid is produced—*a*, from *ammonia*, either that absorbed by the soil from the atmosphere, or that originating in the soil itself by the decay of nitrogenous organic matters. Knop made an experiment with a sandy loam, as follows: The earth was exposed in a box to the vapor of ammonia for three days, was then mixed thoroughly, spread out thinly, moistened with pure water, and kept sheltered from rain until it became dry again. At the beginning of the experiment, 1,000,000 parts of the earth contained 52 parts of nitric acid. During its exposure to the air, while moist, the content of nitric acid in this earth increased to 591 parts in 1,000,000, or more than eleven times; and, as Knop asserts, this increase took place at the expense of the ammonia which the earth had absorbed. The conversion of ammonia into nitric acid is an oxidation expressed by the statement

$$2 NH_3 + 4 O = NH_4 NO_3 + H_2O.\ ^*$$

The oxygen may be either ozone, as already explained, or it may be furnished by a substance which exists in all soils and often to a considerable extent, viz., sesquioxide of iron. This compound ($Fe_2 O_3$) readily yields a portion of its oxygen to bodies which are inclined to oxidize, being itself reduced thereby to protoxide (FeO) thus:— $Fe_2 O_3 = 2 FeO + O$. The protoxide in contact with the air quickly absorbs common oxygen, passing into sesquioxide again, and in this way iron operates as a carrier of atmospheric oxygen to bodies which cannot directly combine with the latter. The oxidizing action of sesquioxide of iron is proved to take place in many instances; for example, a rope tied around a rusty iron bolt becomes "rotten," cotton and linen fabrics are destroyed by iron-stains, the head of an iron nail corrodes away the wood surrounding it, when exposed to the weather, and after suf-

---

* The above equation represents but one-half of the ammonia as converted into nitric acid. In the soil the carbonates of lime, etc., would separate the nitric acid from the remaining ammonia and leave the latter in a condition to be oxidized.

ficient time this oxidation extends so far as to leave the board loose upon the nail, as may often be seen on old, unpainted wooden buildings. Direct experiments by Knop (*Versuchs St.*, III, 228) strongly indicate that ammonia is oxidized by the agency of iron in the soil.

b. The *organic matters* of the soil, either of vegetable or animal origin, *which contain nitrogen*, suffer oxidation by directly combining with ordinary oxygen.

As we shall presently see, nitrates cannot be formed in the rapid or putrefactive stages of decay, but only later, when the process proceeds so slowly that oxygen is in large excess. When the organic matters are so largely diluted or divided by the earthy parts of the soil that oxygen greatly preponderates, it is probable that the nitrogen of the organic bodies is directly oxidized to nitric acid. Otherwise ammonia is first formed, which is converted into nitrates at a subsequent slower stage of decay.

Nitrogenous organic matters may perhaps likewise yield nitric acid when oxidized by the intervention of hydrated sesquioxide of iron, or other reducible mineral compounds. Thenard mentions (*Comptes Rendus*, XLIX, 289) that a nitrogenous substance obtained by him from rotten dung and called *fumic acid*,* when mixed with carbonate of lime, sesquioxide of iron and water, and kept hot for 15 days in a closed vessel, was oxidized with formation of carbonic acid and noticeable quantities of nitric acid, the sesquioxide being at the same time reduced to protoxide.

The various sulphates that occur in soils, especially sulphate of lime (gypsum, plaster), and sulphate of iron (copperas), may not unlikely act in the same manner to convey oxygen to oxidable substances. These sulphates, in exclusion of air, become reduced by organic matters to sulphides. This often happens in deep fissures in the earth, and causes many natural waters to come to the sur-

---

* According to Mulder, impure humate of ammonia.

face charged with sulphides (sulphur-springs). Water containing sulphates in solution often acquires an odor of sulphuretted hydrogen by being kept bottled, the cork or other organic matters deoxidizing the sulphates. The earth just below the paving-stones in Paris contains considerable quantities of sulphides of iron and calcium, the gypsum in the soil being reduced by organic matters. (Chevreul.) These sulphides, when exposed to air, speedily oxidize to sulphates, to suffer reduction again in contact with the appropriate substances, and under certain conditions, operate continuously, to gather and impart oxygen. One of the causes of the often remarkable and inexplicable effects of plaster of Paris when used as a fertilizer may, perhaps, be traced to this power of oxidation, resulting in the formation of nitrates. This point requires and is well worthy of special investigation.

c. Lastly, the free nitrogen of the atmosphere appears to be in some way involved in the act of nitrification—is itself to a certain extent oxidized in the soil, as has been maintained by Saussure, Gaultier de Claubry, and others (*Gmelin's Hand-book of Chemistry,* II, 388).

The truth of this view is sustained by some of Boussingault's researches on the garden soil of Liebfrauenberg. (*Agronomie, etc.,* T., 1, 318). On the 29th of July, 1858, he spread out thinly 120 grammes of this soil in a shallow glass dish, and for three months moistened it daily with water exempt from compounds of nitrogen. At the end of this time analysis of the soil showed that while a small proportion of carbon ($0.825°|_0$) had wasted by oxidation, the quantity of nitrogen had slightly increased. The gain of nitrogen was but 0.009 grm. = $0.008°|_0$.

In five other experiments where plants grew for several months in small quantities of the same garden soil, either in the free air but sheltered from rain and dew, or in a confined space and watered with pure water, analyses were made of the soil and seed before the trial, and of the soil and crop afterwards.

The analyses show that while in all cases the plants gained some nitrogen beyond what was originally contained in the seed, there was in no instance any loss of nitrogen by the soil, and in three cases the soil contained more of this element after than before the trial. Here follow the results.

No. of Exp.	Weight of Crop, the seed taken as 1.	Quantity of Soil.	Gain of Nitrogen	
			by plant.	by soil.
1. Lupin,*	3½	130 grms.	0.0042 grms.	0.0672 grms.
2. Lupin,	4	130 "	0.0047 "	0.0081 "
3. Hemp,	5	40 "	0.0039 "	0.0000 "
4. Bean,	5	50 "	0.0226 "	0.0000 "
5. Lupin,*	3	130 "	0.0217 "	0.0454 "

That the gain of nitrogen by the soil was not due to direct absorption of nitric acid or ammonia from the atmosphere is demonstrated by the fact that it was largest in the two cases (Exps. 1 and 5) where the experiment was conducted in a closed vessel, containing throughout the whole time the same small volume, about 20 gallons, of air.

In Exp. 4, where the soil at the conclusion contained no more nitrogen than at the commencement of the trial, it is scarcely to be doubted that the considerable gain of nitrogen experienced by the plant came through the soil, and would have been found in the latter had it borne no crop.

The experiments show that the quantity of nitrogen assimilated from the atmosphere by a given soil is very variable, or may even amount to nothing (Exp. 3); but they give us no clue to the circumstances or conditions which quantitatively influence the result. It must be observed that this fixation of nitrogen took place here in a soil very rich in organic matters, existing in the condition of humus, and capable of oxidation, so that the soil itself

---

* Experiments made in confined air.

lost during three summer months eight-tenths of one per cent of carbon. In the numerous similar experiments made by Boussingault with soils *destitute of organic matter*, no accumulation of nitrogen occurred beyond the merest traces coming from condensation of atmospheric ammonia.

Certain experiments executed by Mulder more than 20 years ago (*Chemistry of Animal and Vegetable Physiology*, p. 673) confirm the view we have taken. Two of these were "made with beans which had germinated in an atmosphere void of ammonia, and grown, in one case, in ulmic acid prepared from sugar, and also free from ammonia; and, in the other case, in charcoal, both being moistened with distilled water free from ammonia. The ulmic acid and the charcoal were severally mixed up with 1 per cent of wood ashes, to supply the plants with ash-ingredients. I determined the proportion of nitrogen in three beans, and also in the plants that were produced by three other beans. The results are as follows: —

	White beans in ulmic acid.		*Brown beans in charcoal.*	
	Weight.	Nitrogen.	Weight.	Nitrogen.
Beans,	1.465 grm.	50 cub. cent.	1.277	27 cub. cent.
Plants,	4.167 "	160 " "	1.772	54 " "

The white beans, therefore, whilst growing into plants in substances and an atmosphere, both of which were free of ammonia, had obtained more than thrice the quantity of nitrogen that originally existed in the beans; in the brown beans the original quantity was doubled." Mulder believed this experiment to furnish evidence that ammonia is produced by the union of atmospheric nitrogen with hydrogen set free in the decay of organic matters. To this notion allusion has been already made, and the conviction expressed that no proof can be adduced in its favor (p. 239). The results of the experiments are fully explained by assuming that nitrogen was oxidized in nitrification, and no other explanation yet proposed accords with existing facts.

As to the mode in which the soil thus assimilates free nitrogen, several hypotheses have been offered. One is that of Schönbein, to the effect that in the act of evaporation free nitrogen and water combine, with formation of nitrite of ammonia. In a former paragraph, p. 79, we have given the results of Zabelin, which appear to render this theory inadmissible.

A second and adequate explanation is, that free nitrogen existing in the cavities of the soil is directly oxidized to nitric acid by ozone, which is generated in the action of ordinary oxygen on organic matters, (in the same way as happens when ordinary oxygen acts on phosphorus,) or is, perhaps, the result of electrical disturbance.

Experiments by Lawes, Gilbert, and Pugh (*Phil. Trans.*, 1861, II, 495), show indeed that organic matters in certain conditions of decay do not yield nitric acid under the influence of ozone.

They caused air highly impregnated with ozone to pass daily for six months through moist mixtures of burned soil with relatively large quantities of saw-dust, starch, and bean meal, with and without lime—in all 10 mixtures—but in no case was any nitric acid produced.

It would thus appear that ozone can form nitrates in the soil only when organic matters have passed into the comparatively stable condition of humus.

That nitrogen is oxidized in the soil by ozone is highly probable, and in perfect analogy with what must happen in the atmosphere, and is demonstrated to occur in Schönbein's experiments with moistened phosphorus (p. 66, also *Ann. der Chem. u. Pharm.*, 89, 287), as well as in Zabelin's investigations that have been already recounted. (See pp. 75-83.)

he fact, established by Reichardt and Blumtritt, that humus condenses atmospheric nitrogen in its pores (p. 167), doubtless aids the oxidation of this element.

The third mode of accounting for the oxidation of

free nitrogen is based upon the effects of a reducible body, like sesquioxide of iron or sulphate of lime, to which attention has been already directed.

In a very carefully conducted experiment, Cloez* transmitted atmospheric air purified from suspended dust, and from nitric acid and ammonia, through a series of 10 large glass vessels filled with various porous materials. Vessel No. 1 contained fragments of unglazed porcelain; No. 2, calcined pumice-stone; No. 3, bits of well-washed brick. Each of these three vessels also contained 10 grms. of carbonate of potash dissolved in water. The next three vessels, Nos. 4, 5, and 6, included the above-named porous materials in the same order; but instead of carbonate of potash, they were impregnated with carbonate of lime by soaking in water, holding this compound in suspension. The vessel No. 7 was occupied with Mendon chalk, washed and dried. No. 8 contained a clayey soil thoroughly washed with water and ignited so as to carbonize the organic matters without baking the clay. No. 9 held the same earth washed and dried, but not calcined. Lastly, in No. 10, was placed moist pumice-stone mixed with pure carbonate of lime and 10 grms. of urea, the nitrogenous principle of urine. Through these vessels a slow stream of purified air, amounting to 160,000 liters, was passed, night and day, for 8 months. At the conclusion of the experiment, vessel No. 1 contained a minute quantity of nitric acid, which, undoubtedly, came from the atmosphere, having escaped the purifying apparatus. The contents of Nos. 2, 4, and 5, were free from nitrates. Nos. 3 and 6, containing fragments of washed brick, gave notable evidences of nitric acid. Traces were also found in the washed chalk, No. 7, and in the calcined soil, No. 8. In No. 9, filled with washed soil, niter was abundant. No. 10,

---

* Recherches sur la Nitrification—Leçons de Chimie professées en 1861 à la Société Chimique de Paris, pp. 145-150.

containing pumice, carbonate of lime, and urea, was destitute of nitrates.

Experiments 2, 4, and 5, demonstrate that the concourse of nitrogen gas, a porous body, and an alkali-carbonate, is insufficient to produce nitrates. Experiment No. 10 shows that the highly nitrogenous substance, urea,* diffused throughout an extremely porous medium and exposed to the action of the air in moist contact with carbonate of lime, does not suffer nitrification. In the brick (vessels Nos. 3 and 6), something was obviously present, which determined the oxidation of free atmospheric nitrogen. Cloez took the brick fresh from the kiln where it was burned, and assured himself that it included at the beginning of the experiment, no nitrogen in organic combination and no nitrates of any kind. Cloez believes the brick to have contained some oxidable mineral substance, probably sulphide of iron. The Gentilly clay, used in making the brick, as well as some iron-cinder, added to it in the manufacture, furnished the elements of this compound.

The slight nitrification that occurred in the vessels Nos. 7 and 8, containing washed chalk and burned soil, likewise points to the oxidizing action of some mineral matter. In vessel No. 9, the simply washed soil, which was thus freed from nitrates before the trial began, underwent a decided nitrification in remarkable contrast to the same soil calcined (No. 8). The influence of humus is thus brought out in a striking manner.

It may be that apocrenic acid, which readily yields oxygen to oxidable matters, is an important agent in

---

* Urea ($COH_4 N_2$) contains in 100 parts:

Carbon,	20.00
Hydrogen,	6.67
Nitrogen,	46.67
Oxygen,	26.66
	100.00

nitrification. As we have seen, this acid, according to Mulder, passes into crenic acid by loss of oxygen, to be reproduced from the latter by absorption of free oxygen. The apocrenate of sesquioxide of iron, in which both acid and base are susceptible of this transfer of oxygen, should thus exert great oxidizing power. (See p. 228.)

The Conditions Influencing Nitrification have been for the most part already mentioned incidentally. We may, however, advantageously recapitulate them.

a. The formation of nitrates appears to require or to be facilitated by an *elevated temperature*, and goes on most rapidly in hot weather and in hot climates.

b. According to Knop, ammonia that has been absorbed by a soil suffers no change so long as the soil is dry; but when the soil is moistened, nitrification quickly ensues. *Water* thus appears to be indispensable in this process.

c. An *alkali base* or *carbonate* appears to be essential for the nitric acid to combine with. It has been thought that the mere presence of potash, soda, and lime, favors nitrification, "disposes," as is said, nitrogen to unite with oxygen. Boussingault found, however (*Chimie Agricole*, III, 198), that caustic lime developed ammonia from the organic matters of his garden soil without favoring nitrification as much as mere sand. The caustic lime by its chemical action, in fact, opposed nitrification; while pure sand, probably by dividing the particles of earth and thus perfecting their exposure to the air, facilitated this process. Boussingault's experiments on this point were made by inclosing an earth of known composition (from his garden) with sand, etc., in a large glass vessel, and, after three to seven months, analyzing the mixtures, which were made suitably moist at the outset. Below are the results of five experiments.

I. 1000 grms. of soil and 850 grms. sand acquired 0.012 grms. ammonia and 0.482 grms. nitric acid.

II. 1000 grms. of soil and 5500 grms. sand acquired 0.035 grms. ammonia and 0.545 grms. nitric acid.

III. 1000 grms. of soil and 500 grms. marl acquired 0.002 grms. ammonia and 0.360 grms. nitric acid.

IV. 1000 grms. of soil and 2 grms. carbonate of potash acquired 0.015 grms. ammonia and 0.290 grms. nitric acid.

V. 1000 grms. of soil and 200 grms. quicklime acquired 0.303 grms. ammonia and 0.099 grms. nitric acid.

The unfavorable effect of caustic lime is well pronounced and is confirmed by other similar experiments. Carbonate of potash, which is strongly alkaline, but was used in small quantity, and marl (carbonate of lime), which is but very feebly alkaline, are plainly inferior to sand in their influence on the development of nitric acid.

The effect of lime or carbonate of potash in these experiments of Boussingault may, perhaps, be thus explained. Many organic bodies which are comparatively stable of themselves, absorb oxygen with great avidity in presence of, or rather when combined with, a caustic alkali. Crenic acid is of this kind; also gallic acid (derived from nut-galls), and especially pyrogallic acid (a result of the dry distillation of gallic acid). The last-named body, when dissolved in potash, almost instantly removes the oxygen from a limited volume of air, and is hence used for analysis of the atmosphere.*

We reason, then, that certain organic matters in the soil of Boussingault's garden, became so altered by treatment with lime or carbonate of potash as to be susceptible of a rapid oxidation, in a manner analogous to what happens with pyrogallic acid. Dr. R. Angus Smith has shown (*Jour. Roy. Ag. Soc.*, XVII, 436) that if a soil rich in organic matter be made alkaline, moist, and warm, putrefactive decomposition may shortly set in. This is what happens in every well-managed compost of lime and peat. By this rapid alteration of organic matters, as we shall see (p. 268), not only is nitric acid not formed, but nitrates added are reduced to ammonia. It is not improbable that

---

* Not all organic bodies, by any means, are thus affected. Lime *hinders* the alteration of urine, flesh, and the albuminoids.

smaller doses of lime or alkali than those employed by Boussingault would have been found promotive of nitrification, especially after the lapse of time sufficient to allow the first rapid decomposition to subside, for then we should expect that its presence would favor slow oxidation. This view is in accordance with the idea, universally received, that lime, or alkali of some sort, is an indispensable ingredient of artificial niter-beds. The point is one upon which further investigations are needed.

d. Free oxygen, i. e., atmospheric air, and the porosity of soil which ensures its contact with the particles of the latter, are indispensable to nitrification, which is in all cases a process of oxidation. When sesquioxide of iron oxidizes organic matters, its action would cease as soon as its reduction to protoxide is complete, but for the atmospheric oxygen, which at once combines with the protoxide, constantly reproducing the sesquioxide.

In the saltpeter plantations it is a matter of experience that light, porous soils yield the largest product. The operations of tillage, which promote access of air to the deeper portions of earth and counteract the tendency of many soils to "cake" to a comparatively impervious mass, must also favor the formation of nitrates.

Many authors, especially Mulder, insist upon the physical influence of *porosity* in determining nitrification by condensed oxygen. The probability that porosity may assist this process where compounds of nitrogen are concerned, is indeed great; but there is no evidence that any porous body can determine the union of free nitrogen and oxygen. Knop found that of all the proximate ingredients of the soil, clay alone can be shown to be capable of physically condensing gaseous ammonia (humus combines with it chemically, and if it previously effects physical condensation, the fact cannot be demonstrated).

The observations by Reichardt and Blumtritt on the condensing effect of the soil for the gases of the atmosphere (p. 167) indicate absorption both of oxygen and nitrogen, as well as of carbonic acid. The fact that charcoal acts as an energetic oxidizer of organic matters has been alluded to (p. 109). This action is something very remarkable, although charcoal condenses oxygen but to a slight extent. The soil exercises a similar but less vigorous oxidizing effect, as the author is convinced from experiments made under his direction (by J. J. Matthias, Esq.), and as is to be inferred from the well-known fact that the odor of putrefying flesh, etc., cannot pass a certain thickness of soil. But charcoal is unable to accomplish the union of oxygen and nitrogen at common temperatures, or at 212° F., either dry, moistened with pure water, or with solution of caustic soda. (Experiments in Sheffield laboratory, by Dr. L. H. Wood.)

Putrefying flesh, covered with charcoal as in Stenhouse's experiment (p. 169) gives off ammonia, but no nitric acid is formed. Dumas has indeed stated (*Comptes Rend.*, XXIII) that ammonia mixed with air is converted into nitric acid by a porous body—chalk—that has been drenched with caustic potash and is heated to 212° F. But this is an error, as Dr. Wood has demonstrated. It is true that platinum at a high temperature causes ammonia and oxygen to unite. Even a platinum wire when heated to redness exerts this effect in a striking manner (Kraut, *Ann. Ch. u. Ph.*, 136, 69); but spongy platinum is without effect on a mixture of air and ammonia gas at 212° or lower temperatures. (Wood.)

e. Presence of organic matters prone to oxidation. Reduction of nitrates to ammonia, etc., in the soil.—As we have seen, the organic matters (humus) of the soil are a source of nitric acid. But it appears that this is not always or universally true. In compact soils, at a certain depth, organic matters (their hydrogen and carbon) may oxidize at the expense of nitric acid itself, converting the latter into ammonia. Pelouze (*Comptes Rendus*, XLIV,

118) has proved that putrefying animal substances, as albumin, thus reduce nitric acid with formation of ammonia. For this reason, he adds, the liquor of dung heaps and putrid urine contains little or no nitrates. Boussingault (*Agronomie*, II, 17) examined a remarkably rich alluvial soil from the junction of the Amazon with the Rio Cupari, made up of alternate layers of sand and partially decayed leaves, containing $40°|_0$ of the latter. This natural leaf-compost contained no trace of nitrates, but an exceptionally high quantity of ammonia, viz., $.05°|_0$.

Kuhlmann (*Ann. de Chim. et de Phys.*, 3 Ser., XX) was the first to draw attention to the probability that nitric acid may thus be deoxidized in the lower strata of the soil, and his arguments, drawn from facts observed in the laboratory, appear to apply in cases where there exist much organic matters and imperfect access of air. In a soil so porous as is demanded for the culture of most crops these conditions cannot usually occur, as Grouven has taken the trouble to demonstrate (*Zeitschrift für Deutsche Landwirthe*, 1855, p. 341). In rice swamps and peat bogs, as well as in wet compost heaps, this reduction must proceed to a considerable extent.

In some, if not all cases, the addition of much lime or other alkaline substance to a soil rich in organic matters sets up rapid putrefactive decomposition, whereby nitrates are at once reduced to ammonia (p. 266).

In one and the same soil the conditions may exist at different times that favor nitrification on the one hand, and reduction of nitrates to ammonia on the other. A *surplus of moisture* might so exclude air from a porous soil as to cause reduction to take place, to be succeeded by rapid nitrification as the soil becomes more dry.

It is possible that nitrates may undergo further chemical alteration in the presence of excess of organic matters. That nitrites may often exist in the soil is evident from what has been written with regard to the mutual convertibility of nitrates and nitrites (p. 73). According to Goppelsröder (*Dingler's Polytech. Jour.*, 164, 388), certain soils rich in humus possess in a high degree the power to reduce nitrates to nitrites. It is not unlikely that further reduction may occur—that, in fact, the deoxidation may be complete and free nitrogen be disengaged. This is a question eminently worthy of study.

Loss of Nitrates may occur when the soil is saturated with water, so that the latter actually flows through and away from it, as happens during heavy rains, the nitrates (those of sesquioxide of iron, perhaps, excepted) being freely soluble and not retained by the soil. Boussingault made 40 analyses of lake and river water, 25 of spring water, and 35 of well water, and found nitric acid in every case, though the quantity varied greatly, being largest in cities and fertile regions. Thus the water of the upper Rhine contains one millionth, that of the Seine, in Paris, six millionths, and that of the Nile four millionths of nitric acid. The Rhine daily removes from the country supplying its waters an amount of nitric acid equivalent to 220 tons of saltpeter. The Seine carries daily into the Atlantic 270 tons, and the Nile pours 1,100 tons into the Mediterranean every twenty-four hours.

In the wells of crowded cities the proportion of nitrates is much higher. In the older parts of Paris the well waters contain as much as one part of niter (or its equivalent of other nitrates) in 500 of water.

The soil may experience a loss of nitrates by the complete reduction of nitric acid to gaseous nitrogen, or by the formation of inert compounds with humus, as will be noticed in the next section.

Loss of assimilable nitrogen by the washing of nitrates from the soil may be hindered to some extent in compact soils by the fact just noticed that nitric acid is liable to be converted into ammonia, which is at once rendered comparatively insoluble.

Nitric Acid as Food to Plants.—Experiments demonstrating that nitric acid is capable of perfectly supplying vegetation with nitrogen were first made by Boussingault (*Agronomie, Chimie Agricole, etc.*, 1, 210). We give an account of some of these.

Two seeds of a dwarf Sunflower (*Helianthus argophyllus*), were planted in each of three pots, the soil of which, consisting of a mixture of brick-dust and sand, as well as the pots themselves, had been thoroughly freed from all nitrogenous compounds by ignition and washing with distilled water.

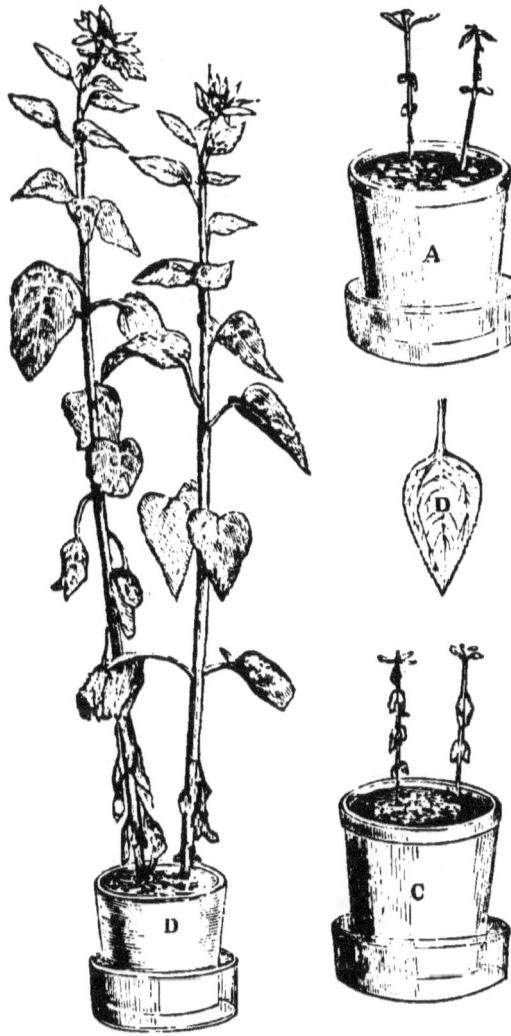

Fig. 9.

To the soil of the pot A, fig. 9, nothing was added save the two seeds, and distilled water, with which all the plants were watered from time to time. With the soil of pot C, were incorporated small quantities of phosphate of lime, of ashes of clover, and bicarbonate of potash, in order that the plants growing in it might have an abundant supply of all the ash-ingredients they needed. Finally, the soil of pot D received the same mineral matters as pot C, and, in addition, a small quantity (1.4 gram) of nitrate of potash. The seeds were sown on the 5th of July, and on the 30th of September, the plants had the relative size and appearance seen in the figure, where they are represented in one-eighth of the natural dimensions.

For the sake of comparison, the size of one of the largest leaves of the same kind of Sunflower that grew in the garden is represented at D, in one-eighth of its natural dimensions.

Nothing can be more striking than the influence of the nitrate on the growth of this plant, as exhibited in this experiment. The plants A and C are mere dwarfs, although both carry small and imperfectly developed flowers. The plant D, on the contrary, is scarcely smaller than the same kind of plant growing under the best conditions of garden culture. Here follows a Table of the results obtained by the examination of the plants.

	Weight of dry crop, the seeds taken as 1.	Vegetable matter organized.	O_2 decomposed in 24 hours.	Acquired by the plants in 86 days of vegetation.	
				Carbon.	Nitrogen.
		grm.	cubic cent.	grm.	grm.
A—nothing added to the soil.........	3.6	0.285	2.45	0.114	0.0023
C—ashes, phosphate of lime, and bicarbonate of potash, added to the soil................	4.6	0.391	3.42	0.156	0.0027
D—ashes, phosphate of lime, and nitrate of potash, added to the soil.	198.3	21.111	182.00	8.444	0.1666

We gather from the above data:

1. That without some compound of nitrogen *in the soil* vegetation cannot attain any considerable development, notwithstanding all requisite ash-ingredients are present

in abundance. Observe that in exps. A and C the crop attained but 4 to 5 times greater weight than the seed, and gathered from the atmosphere during 86 days but 2½ milligrams of nitrogen. The crop, supplied with nitrate of potash, weighed 200 times as much as the seed, and assimilated 66 times as much nitrogen as was acquired by A and C from external sources.

2. That nitric acid of itself may furnish *all the nitrogen* requisite to a normal vegetation.

In another series of experiments (*Agronomie, etc.*, I, pp. 227–233) Boussingault prepared four pots, each containing 145 grams (about 5 oz. avoirdupois) of calcined sand with a little phosphate of lime and ashes of stable-dung, and planted in each two Sunflower seeds. To three of the pots he added weighed quantities of nitrate of soda—to No. 3 twice as much as to No. 2, and to No. 4 three times as much as to No. 3; No. 1 received no nitrate. The seeds germinated duly, and the plants, sheltered from rain and dew, but fully exposed to air, and watered with water exempt from ammonia, grew for 50 days. In the subjoined Table is a summary of the results.

Experiment.	N.* of seed.	N. added as nitrate of soda.	Total N. at disposal of plants.	Total N. of crop.	N. taken from the air, or† left in the soil.†	Vegetable matter organized in 50 days of growth.	Relative weights of matter organized, that of first Exp. taken as unity.	Relative weights of N. at disposal of crops, that of the first Exp. taken as unity.
	grms.	grms.	grms.	grms.	grms.	grms.	grms.	grms.
1..	0.0033	0.0000	0.0033	0.0053	0.0020+	0.397	1	1
2..	0.0033	0.0033	0.0066	0.0063	0.0002‡	0.720	1.8	2
3..	0.0033	0.0066	0.0099	0.0097	0.0002‡	1.130	2.8	3
4..	0.0033	0.0264	0.0297	0.0251	0.0046‡	3.280	8.5	9

* N=Nitrogen.

In the first Exp. a trifling quantity of nitrogen was gathered (as ammonia?) from the air. In the others, and especially in the last, nitrate of soda remained in the soil, not having been absorbed entirely by the plants. Observe, however, what a remarkable coincidence exists between the ratios of supply of nitrogen in form of a nitrate and those of growth of the several crops, as exhibited in the last two columns of the Table. Nothing could demonstrate more strikingly the nutritive function of nitric acid than these admirable investigations.

Of the multitude of experiments on vegetable nutrition which have been recently made by the process of water-culture (*H. C. G.*, p. 167), nearly all have depended upon nitric acid as the exclusive source of nitrogen, and it has proved in all cases not only adequate to this purpose, but far more certain in its effects than ammonia or any other nitrogenous compound.

§ 6.

NITROGENOUS ORGANIC MATTERS OF THE SOIL.
AVAILABLE NITROGEN.—QUANTITY OF NITROGEN REQUIRED FOR CROPS.

In the minerals and rocks of the earth's surface nitrogen is a very small, scarcely appreciable ingredient. So far as we now know, ammonia-salts and nitrates (nitrites) are the only mineral compounds of nitrogen found in soils. When, however, organic matters are altered to humus, and become a part of the soil, its content of nitrogen acquires significance. In peat, which is humus comparatively free from earthy matters, the proportion of nitrogen is often very considerable. In 32 specimens of peat examined by the author (*Peat and its Uses as Fertilizer and Fuel*, p. 90), the nitrogen, *calculated on the organic matters*, ranged from 1.12 to 4.31 per cent, the average being 2.6 per cent. The average amount of nitrogen in the air-dry and in some cases highly impure peat, was 1.4 per cent. This nitrogen belongs to the organic matters in great part, but a small proportion of it being in the form of ammonia-salts or nitrates.

In 1846, Krocker, in Liebig's laboratory, first estimated the nitrogen in a number of soils and marls (*Ann. Ch. u. Ph.*, 58, 387). Ten soils, which were of a clayey or loamy character, yielded from 0.11 to 0.14 per cent; three sands gave from 0.025 to 0.074 per cent; seven marls contained 0.004 to 0.083 per cent.

Numerous examinations have since been made by Anderson, Liebig, Ritthausen, Wolff, and others, with similar results.

In all but his latest writings, Liebig has regarded this nitrogen as available to vegetation, and in fact designated it as ammonia. Way, Wolff, and others, have made evident that a large portion of it exists in organic combination. Boussingault (*Agronomie*, T. I) has investigated the subject most fully, and has shown that in rich and highly manured soils nitrogen accumulates in considerable quantity, but exists for the most part in an insoluble and inert form. In the garden of Liebfrauenberg, which had been heavily manured for centuries, but $4°|_0$ of the total nitrogen existed as ammonia-salts and nitrates. The soil itself contained—

Total nitrogen, 0.261 per cent.
Ammonia, 0.0022 " "
Nitric acid, 0.00034 " "

The subjoined Table includes the results of Boussingault's examinations of a number of soils from France and South America, in which are given the quantities of ammonia, of nitric acid, expressed as nitrate of potash, and of nitrogen in organic combination. These quantities are stated both in per cent of the air-dry soil, and in lbs. av. per acre, taken to the depth of 17 inches. In another column is also given the ratio of nitrogen to carbon in the organic matters. (*Agronomie*, T II, pp. 14-21.)

AMMONIA, NITRATES, AND ORGANIC NITROGEN OF VARIOUS SOILS.

Soils.	Ammonia.		Nitrate of potash.		Nitrogen in org. combi'n.		Ratio of nitrogen to carbon.
	per cent.	lbs. per acre	per cent	lbs. per acre	per cent.	lbs per acre	
South America, France. Liebfrauenberg, light gard. soil	0.0022	100	0.0175*	875	0.259	12970	1:9.3
Bischwiller, light garden soil...	0.0020	100	0.1526	7630	0.295	14755	1:9.7
Bechelbronn, wheat field clay.	0.0009	45	0.0015	75	0.139	6985	1:8.2
Argentan, rich pasture.........	0.0060	300	0.0046	230	0.513	25650	1:8
Rio Madeira, sugar field, clay	0.0090	450	0.0004	20	0.143	7140	1:6.3
Rio Trombetto, forest heavy do.	0.0050	183	0.0001	5	0.119	5955	1:4.9
Rio Negro, prairie v. fine sand.	0.0038	190	0.0001	5	0.068	3410	1:5.6
Santarem, cocoa plantation..	0.0083	415	0.0011	55	0.649	32450	1:11
Saracea, near Amazon, loam..	0.0042	210	none		0.182	9100	1:8.2
Rio Cupari, rich leaf mould....	0.0525	2875	"		0.685	34250	1:18.8
Iles du Salut, French Guiana...	0.0030	400	0.0643	3215	0.543	27170	1:11.7
Martinique, sugar field.........	0.0055	275	0.0186	930	0.112	5590	1:8

* The same soil whose partial analysis has just been given, but examined for nitrates at another time.

It is seen that in all cases the nitrogen in the forms of ammonia † and nitrates ‡ is much less than that in organic combination, and in most cases, as in the Liebfrauenberg garden, the disparity is very great.

Nature of the Nitrogenous Organic Matters. Amides. —Hitherto we have followed Mulder in assuming that the humic, ulmic, crenic, and apocrenic acids, are destitute of nitrogen. Certain it is, however, that natural humus is never destitute of nitrogen, and, as we have remarked in case of peat, contains this element in considerable quantity, often 3 per cent or more. Mulder teaches that the acids of humus, themselves free from nitrogen, are naturally combined to ammonia, but that this ammonia is with difficulty expelled from them, or is indeed impossible to separate completely by the action of solutions of the fixed alkalies. In all chemistry, beside, there is no example of such a deportment, and we may well doubt whether the ammonia that is slowly evolved when natural humus is boiled with potash is thus expelled from a humate of ammonia. It is more accordant with general analogies to

---

† Ammonia contains 82.4 per cent of nitrogen.
‡ Nitrate of potash contains 13.8 per cent of nitrogen.

suppose that it is *generated* by the action of the alkali. In fact, there are a large number of bodies which manifest a similar deportment. Many substances which are produced from ammonia-compounds by heat and otherwise, and called *amides*, to which allusion has been already made, p. 276, are of this kind. Oxalate of ammonia, when heated to decomposition, yields oxamide, which contains the elements of the oxalate *minus* the elements of two molecules of water, viz.,

Oxalate of ammonia.   *Oxamide.*   *Water.*
$$2\,(NH_4)\,C_2O_4 = 2\,(NH_2)\,C_2O_2 + 2\,H_2O$$

On boiling oxamide with solution of potash, ammonia is reproduced by the taking up of two molecules of water, and passes off as a gas, while oxalate of potash remains in the liquid.

Nearly every organic acid known has one or several amides, bearing to it a relation similar to that thus subsisting between oxalic acid and oxamide.

Asparagine, a crystallizable body found in asparagus and many other plants, already mentioned as an amide, is thought to be an amide of malic acid.

Urea, the principal solid ingredient of human urine, is an amide of carbonic acid. Uric acid, hippuric acid, guanine, found also in urine; kreatin and kreatinine, occurring in the juice of flesh; thein, the active principle of tea and coffee; and theobromin, that of chocolate, are all regarded as amides.

Amide-like bodies are gelatine (glue), the organic substance of the tendons and of bones, that of skin, hair, wool, and horn. The albuminoids themselves are amide-like, in so far that they yield ammonia on heating with solutions of caustic alkalies.

Albuminoids a Source of the Nitrogen of Humus.— The organic nitrogen of humus may come from the albuminoids of the vegetation that has decayed upon or in the soil. In their alteration by decay, a portion of nitrogen assumes the gaseous form, but a portion remains in an insoluble and comparatively unalterable condition, though in what particular compounds we are unable to say. The loss of carbon and hydrogen from decaying organic matters, it is believed, usually proceeds more rapidly than the waste of nitrogen, so that in humus, which is the residue of the change, the relative proportion of nitrogen to carbon is greater than in the original vegetation.

Reversion of Nitric Acid and Ammonia to inert Forms.—It is probable that the nitrogen of ammonia, and of nitrates, which are reducible to ammonia under certain conditions, may pass into organic combination in the soil. Knop (*Versuchs St.*, III, 228) found that when peat or soils containing humus were kept for several months in contact with ammonia in closed vessels, at the usual temperature of summer, the ammonia, according to its quantity, completely or in part disappeared. There having been no such amount of oxygen present as would be necessary to convert it into nitric acid, the only explanation is that the ammonia combined with some organic substance in the humus, forming an amide-like body, not decomposable by the hypochlorite of soda used in Knop's azometrical analysis.

Facts supporting the above view by analogy are not wanting. When gelatine (a body of animal origin closely related to the albuminoids, but containing 18 instead of 15°|₀ of nitrogen) is boiled with dilute acids for some time, it yields, among other products, sugar, as Gerhardt has demonstrated. Prof. T. Sterry Hunt was the first to suggest (*Am. Jour. Sci. & Arts*, 1848, Vol. 5, p. 76) that gelatine has nearly the composition of an amide of dextrin or other body of the cellulose group, and might be regarded as derived chemically from dextrin (or starch) by the union of the latter with ammonia, water being eliminated, viz.:

Carbohydrate.   Ammonia.   Water.   Gelatine.
$$C_{12}H_{20}O_{10} + 4\,NH_3 = 6\,H_2O + 2\,(C_6H_{10}N_2O_2).$$
Afterwards Dusart, Schützenberger, and P. Thenard, independently of each other, obtained by exposing dextrin, starch, and glucose, to a somewhat elevated temperature (300–360° F.), in contact with ammonia-water, substances containing from 11 to 19°|₀ of nitrogen, some soluble in water and having properties not unlike those of gelatine, others insoluble. It was observed, also, that analogous compounds, containing less nitrogen, were formed at lower temperatures, as at 212° F. Payen had previously observed that cane sugar underwent entire alteration by prolonged action of ammonia at common temperatures.

These facts scarcely leave room to doubt that ammonia, as carbonate, by prolonged contact with the humic acids or with cellulose, and bodies of like composition, may form combinations with them, from which, by the action of alkalies or lime, ammonia may be regenerated.

It has already been mentioned that when soils are boiled with solutions of potash, they yield ammonia continuously for a long time.

Boussingault observed, as has been previously remarked, that lime, when incorporated with the soil at the ordinary temperature, causes its content of ammonia to increase.

Soil from the Liebfrauenberg garden, mixed with ¹|₈ its weight of lime and nearly ½ its weight of water, was placed in a confined atmosphere for 8 months. On opening the vessel, a distinct odor of ammonia was perceptible, and the earth, which originally contained per kilogram, 11 milligrams of this substance, yielded by analysis 303 mgr. (See p. 265, for other similar results.)

Alteration of Albuminoids in the Soil.—Albuminoids are carried into the soil when fresh vegetable matter is incorporated with it. They are so susceptible to alteration, however, that under ordinary conditions they must speedily decompose, and cannot therefore themselves be considered as ingredients of the soil.

Among the proximate products of their decomposition are organic acids (butyric, valeric, propionic) destitute of nitrogen, and the amides leucin ($C_6H_{13}NO_2$) and tyrosin ($C_9H_{11}NO_3$). These latter bodies, by further decomposition, yield ammonia. As has been remarked, it is probable that the albuminoids, when associated as they are in decay with cellulose and other carbohydrates, may at once give rise to insoluble amide-like bodies, such as those whose existence in humus is evident from the considerations already advanced.

Can these Organic Bodies Yield Nitrogen Directly to Plants ?—Those nitrogenous organic compounds that exist in the soil associated with humus, which possess something of the nature of amides, though unknown to us in a pure state, appear to be nearly or entirely incapable of feeding vegetation directly. Our information on this point is derived from the researches of Boussingault, whose papers on this subject (*De la Terre végétale considérée dans ses effets sur la Végétation*) are to be found in his *Agronomie, etc.*, Vols. I and II.

Boussingault experimented with the extremely fertile soil of his garden, which was rich in all the elements needful to support vegetation, as was demonstrated by the results of actual garden culture. This soil was especially rich in nitrogen, containing of this element $0.26°|_0$, which, were it in the form of ammonia, would be equivalent to more than 7 tons per acre taken to the depth of 13 inches; or, if existing as nitric acid, would correspond to more than 43 tons of saltpeter to the acre taken to the depth just mentioned.

This soil, however, when employed in quantities of 40 to 130 grams (1½ to 4½ oz. av.) and shielded from rain and dew, was scarcely more capable of carrying lupins, beans, maize, or hemp, to any considerable development, than the most barren sand. In eight distinct trials the crops weighed (dry) but 3 to 5 times, in one case 8 times (average 4 times), as much as the seed; while in sand, pumice, or burned soil, containing no nitrogen, Boussingault several times realized a crop weighing 6 times as much as the seed, though the average crop of 38 experiments was but 3 times, and the lowest result 1½ times the weight of the seed.

The fact that the nitrogen of this garden soil was for the most part inert is strikingly shown on a comparison of the crops yielded by it to those obtained in barren soil with aid of known quantities of nitrates.

In a series of experiments with the Sunflower, Boussingault (*Agronomie, etc.*, I, p. 233) obtained in a soil destitute of nitrogen a crop weighing (dry) 4.6 times as much as the seeds, the latter furnishing the plants 0.0033 grm. of nitrogen. In a second pot, with same weight of seeds, in which the nitrogen was doubled by adding 0.0033 grm. in form of nitrate of soda, the weight of crop was nearly doubled—was 7.6 times that of seeds. In a third pot the nitrogen was trebled by adding 0.0066 grm. in form of nitrate, and the crop was nearly trebled also—was 11.3 times the weight of the seeds.

In another experiment (p. 271) the addition of 0.194 grm. of nitrogen as nitrate of potash to barren sand with needful mineral matters, gave a crop weighing 198 times as much as the seeds. But in the garden soil, which contained, when 40 grms. were employed 0.104 grm., and when 130 grms. were used 0.338 grm. of nitrogen, the result of growth was often not greater than in a soil that contained no nitrogen, and only in a single instance surpassed that of a soil to which was added but 0.0033 grm. The fact is thus demonstrated that but a very small proportion of the nitrogen of this soil was assimilable to vegetation.

From these beautiful investigations Boussingault deems it highly probable that in this garden soil, and in soils generally which have not been recently manured, ammonia and nitric acid are the exclusive feeders of vegetation with nitrogen. Such a view is not indeed absolutely demonstrated, but the experiments alluded to render it in the highest degree probable, and justify us in designating the organic nitrogen for the most part as inert, so far as vegetable nutrition is concerned, until altered to nitrates or ammonia-salts by chemical change.

To comprehend the favorable results of garden-culture in such a soil, it must be considered what a large quantity of earth is at the disposal of the crop, viz., as Boussingault ascertained, 57 lbs. for each hill of dwarf beans, 190 lbs. for each hill of potatoes, 470 lbs. for each tobacco plant, and 2,900 lbs. for every three hop-plants.

The quantity and condition of the nitrogen of Boussingault's garden soil are stated in the subjoined scheme.

Available nitrogen { Ammonia 0.00220 per cent = Nitrogen 0.00181 per cent } 0.0019 per ct.
{ Nitric acid 0.00034 "   " =   "   0.00009 "   " }
Inert nitrogen—of organic compounds..............................0.2591 " "

Total nitrogen.................... .... 0.2610 per ct.

Calculation shows that in garden culture the plants above named would have at their disposal in this soil quantities of inert and available nitrogen as follows:

	Weight of soil.	Inert nitrogen.	Available nitrogen.
Bean (dwarf) hill	57 lbs.	75 grams.*	1 gram.
Potato, "	190 "	212 "	3 grams.
Tobacco, single plant,	470 "	555 "	7 "
Hop, three plants,	2900 "	3438 "	44 "

* 1 gram = 15 grains avoirdupois nearly.
17 grams = 1 oz. " "
283 " = 1 lb. " "

Indirect Feeding of Crops by the Organic Nitrogen of the Soil.—In what has been said of the oxidation of the organic matters of the soil, (whereby it is probable that their nitrogen is partially converted into nitric acid,) and of the effect of alkalies and lime upon them, (whereby ammonia is generated,) is given a clue to the understanding of their indirect nutritive influence upon vegetation. By these chemical transformations the organic nitrogen may pass into the two compounds which, in the present state of knowledge, we must regard as practically the exclusive feeders of the plant with nitrogen. The rapidity and completeness of the transformation depend upon circumstances or conditions which we understand but imperfectly, and which are extremely important subjects for further investigation.

Difficulty of estimating the Available Nitrogen of any Soil.—The value of a soil as to its power of supplying plants with nitrogen is a problem by no means easy to solve. The calculations that have just been made from the analytical data of Boussingault regarding the soil of his garden are necessarily based on the assumption that no alteration in the condition of the nitrogen could take place during the period of growth. In reality, however, there is no constancy either in the absolute quantity of nitrogen in the soil or in its state of availability. Portions of nitrogen, both from the air and from fertilizers, may continually enter the soil and assume temporarily the form of insoluble and inert organic combinations. Other portions, again, at the same time and as continually, may escape from this condition and be washed out or gathered by vegetation in the form of soluble nitrates, as has already been set forth. It is then manifestly impossible to learn more from analysis, than how much nitrogen is available to vegetation *at the moment the sample is examined.* To estimate with accuracy what is assimilable during the whole season of growth is simply out of the question. The nearest approach that can be made to this result is to ascertain how much a crop can gather from a limited volume of the soil.

Bretschneider's Experiments.—We may introduce here a notice of some recent researches made by Bretschneider in Silesia, a brief account of which has appeared since the foregoing paragraphs were written. (*Jahresbericht ü. Ag. Chem.*, 1865, 29.)

Bretschneider's experiments were made for the purpose of estimating how much ammonia, nitric acid, and nitrogen, exist or are formed in the soil, either fallow or occupied with various crops during the period of growth. For this purpose he measured off in the field four plots of ground, each one square rod (Prussian) in area, and separated from the others by paths a yard wide. The soil of one plot was dug out to the depth of 12 inches, sifted, and after a board frame 12 inches deep had been fitted to the sides of the excavation, the sifted earth was filled in again. This and another—not sifted—plot were planted to sugar beets, another was sown to vetches, and the fourth to oats.

At the end of April, six accurate and concordant analyses were made of the soil. Afterwards, at five different periods, a cubic foot of soil was taken from each plot, and from the spaces between that bore no vegetation, for determining the amounts of nitric acid, ammonia, and total nitrogen. The results of this analytical work are given in the following Tables, being calculated in pounds for the area of an acre, and to the depth of 12 inches (English measures*):

TABLE I.
AMOUNT OF AMMONIA.

	Beet plot, sifted soil.	Beet plot.	Vetch plot.	Oat plot.	Vacant plot.
End of April,	59	59	59	59	59
12th June,	15	48	41	32	28
30th June,	12	41	24	40	32
22d July,	9	29	39	22	29
13th August,	8	15	16	11	43
9th September,	0	16	16	7	23

---

* It is plain that when the results of analyses made on a small amount of soil are calculated upon the 3,500,000 lbs. of soil (more or less) contained in an acre to the depth of one foot (see p. 158), the errors of the analyses, which cannot be absolutely exact, are enormously multiplied. What allowance ought to be made in this case we cannot say, but should suppose that 5 per cent would not be too much. On this basis differences of 200-300 lbs. in Table IV should be overlooked.

TABLE II.
AMOUNT OF NITRIC ACID.

	Beet plot, sifted soil.	Beet plot.	Vetch plot.	Oat plot.	Vacant plot.
End of April,	56	56	56	56	56
12th June,	281	270	102	28	106
30th June,	328	442	15	93	318
22d July,	116	89	58	0	43
13th August,	53	6	71	14	81
9th September.	0	0	12	0	0

TABLE III.
TOTAL ASSIMILABLE NITROGEN (OF AMMONIA AND NITRIC ACID).

	Beet plot, sifted soil.	Beet plot.	Vetch plot.	Oat plot.	Vacant plot.
End of April,	63	63	63	63	63
12th June,	84	109	60	33	50
30th June,	95	148	23	57	108
22d July,	37	47	31	18	35
13th August,	21	14	31	13	56
9th September,	0	13	16	6	19

TABLE IV.
TOTAL NITROGEN OF THE SOIL.

	Beet plot, sifted soil.	Beet plot.	Vetch plot.	Oat plot.	Vacant plot.
End of April,	4652	4652	4652	4652	4652
12th June,	4861	5209	5606	6140	4720
30th June,	4667	5744	5638	5514	4482
22d July,	5308	5485		4724	4924
13th August,	5467	6316	6316	6266	4412
9th September,	5164	4656	6522	5004	4294

From the first Table we gather that the quantity of *ammonia*, which was considerable in the spring, diminished, especially in a porous (sifted) soil until September. In the compact earth of the uncultivated path, its diminution was less rapid and less complete. The amount of *nitric acid* (nitrates), on the other hand, increased, though not alike in any two cases. It attained its maximum in the hot weather of June, and thence fell off until, at the close of the experiments, it was completely wanting save in a single instance.

The figures in the second Table do not represent the absolute quantities of nitric acid that existed in the soil throughout the period of experiment, but only those amounts that *remained* at the time of taking the samples. What the vegetation took up from the planted plots, what was washed out of the surface soil by rains, or otherwise removed by chemical change, does not come into the reckoning.

Those plots, the surface soil of which was most occupied by active roots, would naturally lose the most nitrates by the agency of vegetation; hence, not unlikely, the vetch and oat plots contained so little in June. The results upon the beet, and vacant ground plots demonstrate that in that month a rapid *formation of nitrates* took place. It is not, perhaps, impossible that nitrification also proceeded vigorously in the loose soils in July and August, but was not revealed by the analysis, either because the vegetation took it up or heavy rains washed it out from the surface soil. In the brief account of these experiments at hand, no information is furnished on these points. Since *moisture* is essential to nitrification, it is possible that a period of dry weather coming on shortly before the soil was analyzed in July, August, and September, had an influence on the results. It is certainly remarkable that with the exception of the vetch plot, the soil was destitute of nitrates on the 9th of September. This plot, at that time, was thickly covered with fallen leaves.

We observe further that the nature of the crops influenced the accumulation of nitrates, whether simply because of the different amount of absorbent rootlets produced by them and unequally developed at the given period, or for other reasons, we cannot decide.*

From the third Table may be gathered some idea of the total quantity of nitrogen that was present in the soil in

---

* It is remarkable that the large-leaved beet plant had a great surplus of nitrates, while the oat plot was comparatively deficient in them. Has this fact any connection with what has been stated (p. 84) regarding the unequal power of plants to provide themselves with nitrogenous food?

a form available to crops. Assuming that ammonia and nitric acid chiefly, if not exclusively, supply vegetation with nitrogen, it is seen that the greatest quantity of available nitrogen ascertained to be present at any time in the soil was 148 lbs. per acre, taken to the depth of one foot. This, as regards nitrogen, corresponds to the following dressings:—

	lbs. per acre.
Saltpeter (nitrate of potash)	1068
Chili saltpeter (nitrate of soda)	898
Sulphate of ammonia	909
Peruvian guano (14 per cent of nitrogen)	1057

The experience of British farmers, among whom all the substances above mentioned have been employed, being that 2 to 3 cwt. of any one of them make a large, and 5 cwt. a very large, application per acre, it is plain that in the surface soil of Bretschneider's trials *there was formed during the growing season a large manuring of nitrates in addition to what was actually consumed by the crops.*

The assimilable nitrogen increased in the beet plots up to the 30th of June, thence rapidly diminished as it did in the soil of the paths. In the oat and vetch plots the soil contained, at none of the times of analysis, so much assimilable nitrogen as at the beginning of the experiments. In September, all the plots were much poorer in available nitrogen than in the spring.

Table IV confirms what Boussingault has taught as to the vast stores of nitrogen which may exist in the soil. The amount here is more than *two tons* per acre. We observe further that in none of the *cultivated plots* did this amount at any time fall below this figure; on the other hand, in most cases it was considerably increased during the period of experiment. In the uncultivated plot, perhaps, the total nitrogen fell off somewhat. This difference may have been due to the root fibrils that, in spite of the utmost care, unavoidably remain in a soil from which growing vegetation is removed. The regular and great increase of total nitrogen in the vetch plot was certainly due in part to the abundance of leaves that fell from the plants, and covered the surface of the soil. But this nitrogen, as well as that of the standing crops, must have come from the atmosphere, since the soil exhibited no diminution in its content of this element.

We have here confirmation of the view that *ammonia*, as *naturally supplied*, is of very trifling importance to vegetation, and that, consequently, *nitrates* are the chief natural means of providing nitrogen for crops. The fact that atmospheric nitrogen becomes a part of the soil and enters speedily into organic and inert combinations, also appears to be sustained by these researches.

Quantity of Nitrogen needful for Maximum Grain Crops.—Hellriegel has made experiments on the effects of various quantities of nitrogen (in the form of nitrates) on the yield of cereals. The plants grew in an artificial soil consisting of pure quartz sand, with an admixture of ash-ingredients in such proportions as trial had demonstrated to be appropriate. All the conditions of the experiments were made as nearly alike as possible, except as regards the amount of nitrogen, which, in a series of eight trials, ranged from nothing to 84 parts per 1,000,000 of soil.

The subjoined Table contains his results.

EFFECTS OF VARIOUS PROPORTIONS OF ASSIMILABLE NITROGEN IN THE SOIL.

Nitrogen in 1,000,000 lbs. of soil.	Yield of Grain, in lbs.					
	Wheat.		Rye.		Oats.	
	Found	Calculated	Found	Calculated	Found	Calculated
0	0.002	—	0.218	—	0.330	—
	Increase		Increase		Increase	
7	0.553	0.926	0.832	0.966	0.929	1.168
14	1.708	1.851	1.911	1.933	2.605	2.336
21	2.767	2.777	2.669	2.899	3.845	3.503
28	3.763	3.703	4.172	3.866	6.211	4.671
42	6.065	5.554	5.162	5.798	7.030	7.007
56	7.198	7.406	7.163	7.732	9.052	9.342
84	9.257	9.257	8.698	8.698	9.342	9.342

From numerous other experiments, not published at this writing, Hellriegel believes himself justified in assuming that the highest yield thus observed, with 84 lbs. of nitrogen in 1,000,000 of soil, might have been got with 70 lbs. of nitrogen in case of wheat, with 63 lbs. in case of rye, and with 56 lbs. in case of oats. On this assumption he has *calculated* the yield of each of these crops, and the figures obtained (see Table) present on the whole a remarkable coincidence with those directly observed.

§ 7.

DECAY OF NITROGENOUS BODIES.

We have incidentally noticed some of the products of the decay of nitrogenous bodies, viz., those which remain in the soil. We may now, with advantage, review the subject connectedly, and make our account of this process more complete.

It will be needful in the first place to give some explanations concerning the nature of the familiar transformations to which animal and vegetable matters are subject.

By the word decay, as popularly employed, is understood a series of chemical changes which are very different in their manifestations and results, according to the circumstances under which they take place or the kinds of matter they attack. Under one set of conditions we have slow decay, or, as Liebig has fitly designated it, *eremeausis;** under others *fermentation;* and under still others *putrefaction.*

Eremecausis* is a slow oxidation, and requires the constant presence of an excess of free oxygen. It proceeds upon vegetable matters which are comparatively

---

* From the Greek, signifying *slow combustion.*

difficult of alteration, such as stems and leaves, consisting chiefly of cellulose, with but little albuminoids, and both in insoluble forms.

What is said in a former paragraph on the "Decay of Vegetation," p. 137, applies in general to cremecausis.

Fermentation is a term commonly applied to any seemingly spontaneous change taking place with vegetable or animal matters, wherein their sensible qualities suffer alteration, and heat becomes perceptible, or gas is rapidly evolved. Chemically speaking, fermentation is the breaking up of an organic body by chemical decomposition, which may go on in absence of oxygen, and is excited by a substance or an organism called a *ferment*.

_{There are a variety of fermentations, viz., the *vinous*, *acetic*, *lactic*, etc. In vinous fermentation, the yeast-fungus, *Torula cerevisiæ*, vegetates in an impure solution of sugar, and causes the latter to break up into alcohol and carbonic acid with small quantities of other products. In the acetic fermentation, the vinegar-plant, *Mycoderma vini*, is believed to facilitate the conversion of alcohol into acetic acid, but this change is also accomplished by platinum sponge, which acts as a ferment. In the lactic fermentation, a fungus, *Penicillium glaucum*, is thought to determine the conversion of sugar into lactic acid, as in the souring of milk.}

_{The transformation of starch into sugar has been termed the saccharous fermentation, diastase being the ferment.}

Putrefaction, or putrid fermentation, is a rapid internal change which proceeds in comparative absence of oxygen. It most readily attacks animal matters which are rich in albuminoids and other nitrogenous and sulphurized principles, as flesh, blood, and urine, or the highly nitrogenous parts of plants, as seeds, when they are fully saturated with water. Putrefying matters commonly disengage stinking gases. According to Pasteur putrefaction is occasioned by the growth of animalcules (*Vibrios*).

Fermentation is usually and putrefaction is always a reducing (deoxidizing) process, for either the ferment itself or the decomposing substances, or some of the products of decomposition, are highly prone to oxidation, and in absence of free oxygen may remove this element from reducible bodies (Traube, *Fermentwirkungen*, pp. 63–78).

In a mixture of cellulose, sugar, and albuminoids, cremecausis, fermentation, and putrefaction, may all proceed simultaneously.

When the albuminoids decay in the soil associated with carbohydrates and humus, the final results of their alteration may be summed up as follows:

1. *Carbon* unites mainly with oxygen, forming carbonic acid gas, which escapes into the atmosphere. With imperfect supplies of oxygen, as when submerged in water, carbonic oxide (CO) and marsh gas (CH_4) are formed. A portion of carbon remains as humus.

2. *Hydrogen*, for the most part, combines with oxygen, yielding water. In deficiency of oxygen, some hydrogen escapes as a carbon compound (marsh-gas), or in the free state. If humus remains, hydrogen is one of its constituents.

3. *a. Nitrogen* always unites to a large extent with hydrogen, giving ammonia, which escapes as gaseous carbonate in considerable quantity, unless from presence of carbohydrates much humus is formed, in which case it may be nearly or entirely retained by the latter. Lawes, Gilbert, and Pugh, (*Phil. Trans.* 1861, II., p. 501) made observations on the decay of wheat, barley, and bean seeds, either entire or in form of meal, mixed with a large quantity of soil or powdered pumice, and exposed in various conditions of moisture to a current of air for six months. They found in nine experiments that from 11 to $58°|_0$ of the nitrogen was converted into ammonia, although but a trifling proportion of this (on the average but $0.4°|_0$) escaped in the gaseous form.

b. In presence of excess of oxygen, a portion of nitrogen usually escapes in the *free state*. Reiset proved the escape of free nitrogen from fermenting dung. Boussingault, in his investigations on the assimilability of free nitrogen, found in various vegetation-experiments, in which crushed seeds were used as fertilizers, that nitrogen was lost by assuming some gaseous form. This loss probably took place to some slight extent as ammonia, but chiefly as free nitrogen. Lawes, Gilbert, and Pugh, found in thirteen out of fifteen trials, including the experiments just referred to, that a loss of free nitrogen took place, ranging from 2 to 40 per cent of the total quantity contained originally in the vegetable matters submitted to decomposition. In six experiments the loss was 12 to 13 per cent. In the two cases where no loss of nitrogen occurred, nothing in the circumstances of decay was discoverable to which such exceptional results could be attributed. Other experiments (*Phil. Trans.* 1861, II., p. 509) demonstrated that in *absence of oxygen* no nitrogen was evolved in the free state.

c. Nitric acid is not formed from the nitrogen of organic bodies in rapid or putrefactive decay, but only in slow oxidation or eremecausis of humified matters. Pelouze found no nitrates in the liquor of dung heaps. Lawes, Gilbert, and Pugh, (*loc. cit.*), found no nitric acid when the seed-grains decayed in ordinary air, nor was it produced when ozonized air was passed over moist bean-meal, either alone or mixed with burned soil or with slaked lime, the experiments lasting several months. It thus appears that the carbon and hydrogen of organic matters have such an affinity for oxygen as to prevent the nitrogen from acquiring it in the quicker stages of decay. More than this, as Pelouze has shown (*Comptes Rendus*, XLIV., p. 118), putrefying matters rob nitric acid of its oxygen and convert it into ammonia. We have already remarked that putrefaction and fermentation are reducing processes, and until they have run their course and the organic matters have passed into the comparatively stable forms of humus, their nitrogen appears to be incapable of oxidation. So soon as compounds of carbon and hydrogen are formed, which unite but slowly with free oxygen, so that the latter easily maintains itself in excess, then and not before, the nitrogen begins to combine with oxygen.

4. Finally, the *sulphur* of the albuminoids may be at first partially dissipated as sulphuretted hydrogen gas, while in the slower stages of decay, it is oxidized to sulphuric acid, which remains as sulphates in the soil.

§ 8.

THE NITROGENOUS PRINCIPLES OF URINE.

The question " How Crops Feed " is not fully answered as regards the element Nitrogen, without a consideration of certain substances—ingredients of urine—which may become incorporated with the soil in the use of animal manures.

Professor Way, in his investigation on the "Power of Soils to Absorb Manure," describes the following remarkable experiment: "Three quantities of fresh urine, of 2,000 grains each, were measured out into similar glasses. With one portion its own weight of *sand* was mixed; with another, its own weight of white *clay;* the third being left without admixture of any kind. When smelt immediately after mixture, the sand appeared to have had no effect, whilst the clay mixture had entirely lost the smell of urine. The three glasses were covered lightly with paper and put in a warm place, being examined from time to time. In a few hours it was found that the urine containing sand had become slightly putrid; then followed the natural urine; but the quantity with which clay had been mixed *did not become putrid at all*, and at the end of seven or eight weeks it had only the peculiar smell of fresh urine, without the slightest putridity. The surface of the clay, however, became afterwards covered with a luxuriant growth of confervæ, which did not happen in the other glasses." (*Jour. Roy. Ag. Soc. of Eng.*, XI., 366.)

Professor Way likewise found that filtering urine through clay or simply shaking the two together, allowing the liquid to clear itself, and pouring it off, sufficed to prevent putrefaction, and keep the urine as if fresh for a month or more. Cloez found, as stated on p. 264, that in a mixture of moistened pumice-stone, carbonate of lime, and urea (the nitrogenous principle of urine), no nitrates were formed during eight months' exposure to a slow current of air.

These facts make it necessary to consider in what state the nitrogen of urine is absorbed and assimilated by vegetation.

Urine contains a number of compounds rich in nitrogen, being derived from the waste of the food and tissues of the animal, which require a brief notice.

Urea ($CO\ N_2H_4$)[*] may be obtained from the urine of man as a white crystalline mass or in distinct transparent rhombic crystals, which remain indefinitely unaltered in dry air, and have a cooling, bitterish taste like saltpeter. It is a weak base, and chemists have prepared its nitrate, oxalate, phosphate, etc.

Urea constitutes 2 to 3 per cent of healthy human urine, and a full-grown and robust man excretes of it about 40 grams, or $1^1/_3$ oz. av. daily.

When urine is left to itself, it shortly emits a putrid odor; after a few days or hours the urea it contained entirely disappears, and the liquid smells powerfully of ammonia. Urea, when in contact with the animal matters

---

[*] Carbon............20.00
Hydrogen.......... 6.67
Nitrogen..........46.67
Oxygen............26.66
         100.00

of urine, suffers decomposition, and its elements, combining with the elements of water, are completely transformed into carbonate of ammonia.

Urea.   Water.   Carbonate of Ammonia.
$$CO\,N_2H_4 + 2H_2O = 2(NH_3),\ H_2O,CO_2.$$

As we have learned from Way's experiments, clay is able to remove from urine the "ferment" which occasions its putrefaction.

Urea is abundant in the urine of all carnivorous and herbivorous mammals, and exists in small quantity in the urine of carnivorous birds, but has not been detected in that of herbivorous birds.

Uric acid ($C_5H_4N_4O_3$)* is always present in healthy human urine, but in very minute quantity. It is the chief solid ingredient of the urine of birds and reptiles. Here it exists mainly as urate of ammonia.** The urine of birds and serpents is expelled from the intestine as a white, thickish liquid, which dries to a chalk-like mass. From this, uric acid may be obtained in the form of a white powder, which, when magnified, is seen to consist of minute crystals. By powerful oxidizing agents uric acid is converted into oxalate and carbonate of ammonia, and urea. Peruvian guano, when of good quality, contains some 10 per cent of urate of ammonia.

Hippuric acid ($C_9H_9NO_3$)† is commonly abundant in the urine of the ox, horse, and other herbivorous animals. By boiling down fresh urine of the pastured or hay-fed cow to $^1\!/_6$ its bulk, and adding hydrochloric acid, hippuric acid crystallizes out on cooling in four-sided prisms, often two or three inches in length.

---

* Carbon 35.72	** Carbon 32.43	† Carbon 60.74
Hydrogen 2.38	Hydrogen 3.78	Hydrogen 4.96
Nitrogen 33.33	Nitrogen 37.84	Nitrogen 7.82
Oxygen 28.57	Oxygen 25.95	Oxygen 26.48
100.00	100.00	100.00

Glycocoll or **Glycine*** is a sweet substance that results from the decomposition of hippuric acid under the influence of various agents. It is also a product of the action of acids on gelatine and horn.

Guanine $(C_5H_5N_5O)$ † occurs to the extent of about $1\frac{1}{2}$ per cent in Peruvian guano, and is an ingredient of the liver and pancreas of animals, whence it passes into the excrement in case of birds and spiders. By oxidation it yields among other products urea and oxalic acid.

Kreatin $(C_4H_9N_3O_2)$ ‡ is an organic base existing in very minute quantity in the flesh of animals, and occasionally found in urine.

Cameron was the first, in 1857, to investigate the assimilability of urinary products by vegetation. His experiments (*Chemistry of Agriculture*, pp. 139–144) were made with barley, which was sown in an artificial soil, destitute of nitrogen. Of four pots one remained without a supply of nitrogen, another was manured with sulphate of ammonia, and two received a solution of urea. The pot without nitrogen gave plants 8 inches high, but these developed no seeds. The pot with sulphate of ammonia gave plants 22 inches high, and 300 seeds. Those with urea gave respectively stalks of 26 and 29 inches height, and 252 and 270 seeds. The soil in neither case contained ammonia, the usual decomposition-product of urea. Dr. Cameron justly concluded that urea enters plants unchanged, is assimilated by them, and equals ammonia-salts as a means of supplying nitrogen to vegetation.

The next studies in this direction were made by the author in 1861 (*Am. Jour. Science*, XLI., 27). Experiments were conducted with uric acid, hippuric acid, and guanine.

* Carbon	39.73	† Carbon	32.00	‡ Carbon	36.64
Hydrogen	3.31	Hydrogen	6.67	Hydrogen	6.87
Nitrogen	46.36	Nitrogen	18.67	Nitrogen	32.06
Oxygen	10.60	Oxygen	42.66	Oxygen	24.43
	100.00		100.00		100.00

Washed and ignited flower-pots were employed, to contain, for each trial, a soil consisting of 700 grms. of ignited and washed granitic sand, mixed with 0.25 grm. sulphate of lime, 2 grms. ashes of hay, prepared in a muffle, and 2.75 grms. bone-ashes. This soil was placed upon 100 grms. of clean gravel to serve as drainage.

In each of four pots containing the above soil was deposited, July 6th, a weighed kernel of maize. The pots were watered with equal quantities of distilled water containing a scarcely appreciable trace of ammonia. The seeds germinated in a healthy manner, the plants developed slowly and alike until July 28th, when the addition of nitrogenous matters was begun.

To No. 1, no solid addition was made.

To No. 2 was added, July 28th, 0.420 grm. uric acid.

To No. 3 was added 1.790 grm. hippuric acid, at four different times, viz: July 28, 0.358 grm., Aug. 26th, 0.358 grm., Sept. 16th, 0.716 grm., Oct. 3d, 0.358 grm.

To No. 4 was added 0.4110 grm. hydrochlorate of guanine, viz: July 28th, 0.0822 grm., Aug. 26th, 0.0822 grm., Sept. 16th, 0.1644 grm., Oct. 3d, 0.0822 grm.

The nitrogenous additions contained in each case, 0.140 grm. of nitrogen, and were strewn, as fine powder, over the surface of the soil.

The plants continued to grow or to remain healthy (the lower leaves withering more or less) until they were removed from the soil, Nov. 8th.

The plants exhibited striking differences in their development. No. 1 (no added nitrogen) produced in all seven slender leaves, and attained a height of 7 inches. At the close of the experiment, only the two newest leaves were perfectly fresh; the next was withered and dead throughout one-third of its length. The newer portions of this plant grew chiefly at the expense of the older parts. No sign of floral organs appeared.

No. 2, fed with uric acid, was the best developed plant of the series. At the conclusion of the experiment, it bore ten vigorous leaves, six of which were fresh, and two but partly withered. It was 14 inches high, and carried two rudimentary ears (pistillate flowers), from the upper one of which hung tassels 6 inches long.

No. 3, supplied with hippuric acid, bore eight leaves, four of which were withered, and two rudimentary ears, one of which tasseled. Height, 12 inches.

No. 4, with hydrochlorate of guanine, had six leaves, one withered, and two ears, one of which was tasseled. Height, 12 inches. The weight of the crops (dried at 212° F.), exclusive of the fine rootlets that could not be removed from the soil, was ascertained, with the subjoined results.

	1 Without Nitrogen.	2 Uric Acid.	3 Hippuric Acid.	4 Guanine.
Weight of dried crop,	0.1925 grm.	1.9470 grm.	1.0149 grm.	0.9820 grm.
" " seed,	0.1644 "	.1725 "	0.1752 "	0.1698 "
gain,	0.0291 "	1.7745 "	0.8397 "	0.8122 "

We thus have proof that all the substances employed contributed nitrogen to the growing plant. This is conclusively shown by the fact that the development of pistillate organs, which are especially rich in nitrogen, occurred in the three plants fed with nitrogenous compounds, but was totally wanting in the other. The relation of matter, new-organized by growth, to that derived from the seed, is strikingly seen from a comparison of the ratios of the weight of the seed to the increase of organized matter, the former being taken as unity.

The ratio is approximatively

for No. 1,   1 : 0.2
"   "   2,   1 : 10.2
"   "   3,   1 : 4.8
"   "   4,   1 : 4.8

The relative gain by growth, that of No. 1 assumed as unity, is for No. 1, — 1
" " 2, — 61
" " 3, — 29
" " 4, — 28

The crops were small, principally because the supply of nitrogen was very limited.

These experiments demonstrate that the substances added, in every case, aided growth by supplying nitrogen. They do not, indeed, prove that the organic fertilizers entered as such into the crop without decomposition, but if urea escapes decomposition in a soil, as Cameron and Cloez have shown is true, it is not to be anticipated that the bodies employed in these trials should suffer alteration to ammonia-salts or nitrates.

Hampe afterwards experimented with urea and uric acid by the method of Water-Culture (*Vs. St.*, VII., 308; VIII., 225; IX., 49; and X., 175). He succeeded in producing, by help of urea, maize plants as large as those growing in garden soil, and fully confirmed Cameron's conclusion regarding the assimilability of this substance. Hampe demonstrated that urea entered as such into the plant. In fact, he separated it, in the pure state, from the stems and leaves of the maize which had been produced with its aid.

Hampe's experiments with uric acid in solution showed that this body supplied nitrogen without first assuming the form of ammonia-salts, but it suffered partially if not entirely a decomposition, the nature of which was not determined. Uric acid itself could not be found in the crop.

Hampe's results with hippuric acid were to the effect that this substance furnishes nitrogen without reversion to ammonia, but is resolved into other bodies, probably benzoic acid and glycocoll, which are formed when hippuric acid is subjected to the action of strong acids or ferments.

Hampe, therefore, experimented with glycocoll, and from his trials formed the opinion that this body is directly nutritive. In fact, he obtained with it a crop equal to that yielded by ammonia-salts.

Knop, who made, in 1857, an unsuccessful experiment with hippuric acid, found, in 1866, that glycocoll is assimilated (*Chem. Centralblatt*, 1866, p. 774).

In 1868, Wagner experimented anew with hippuric acid and glycocoll. His results confirm those of Hampe. Wagner, however, deems it probable that hippuric acid enters the plant as such, and is decomposed within it into benzoic acid and glycocoll (*Vs. St.*, XI., p. 294).

Wagner found, also, that kreatin is assimilated by vegetation.

The grand result of these researches is, that the nitrogenous (amide-like) acids and bases which are thrown off in the urinary excretions of animals need not revert, by decay or putrefaction, to inorganic bodies (ammonia or nitric acid), in order to nourish vegetation, but are either immediately, or after undergoing a slight and easy alteration, taken up and assimilated by growing plants.

As a practical result, these facts show that it is not necessary that urine should be fermented before using it as a fertilizer.

§ 9.

COMPARATIVE NUTRITIVE VALUE OF AMMONIA-SALTS AND NITRATES.

The evidence that both ammonia and nitric acid are capable of supplying nitrogen to plants has been set forth. It has been shown further that nitric acid alone can perfectly satisfy the wants of vegetation as regards the element nitrogen. In respect to ammonia, the case has not been similarly made out. We have learned that ammonia occurs, naturally, in too small proportion, either in the atmosphere or the soil, to supply much nitrogen to crops. In exceptional cases, however, as in the leaf-mold of Rio Cupari, examined by Boussingault, p. 276, as well as in lands manured with fermenting dung, or with sulphate or muriate of ammonia, this substance acquires importance from its quantity.

On the assumption that it is the nitrogen of these substances, and not their hydrogen or oxygen, which is of value to the plant, we should anticipate that 17 parts of ammonia would equal 54 parts of nitric acid in nutritive effect, since each of these quantities represents the same amount (14 parts) of nitrogen. The ease with which ammonia and nitric acid are mutually transformed favors this view, but the facts of experience in the actual feeding of vegetation do not, as yet, admit of its acceptance.

In earlier vegetation-experiments, wherein the nitrogenous part of an artificial soil (without humus or clay) consisted of ammonia-salts, it was found that these were decidedly inferior to nitrates in their producing power. This was observed by Ville in trials made with wheat planted in calcined sand, to which was added a given quantity of nitrogen in the several forms of nitrate of potash, sal-ammoniac (chloride of ammonium), nitrate of ammonia, and phosphate of ammonia.

Ville's results are detailed in the following table. The quantity of nitrogen added was 0.110 grm. in each case.

Source of Nitrogen.		Straw and Roots.	Grain.	Average crop.	Nitrogen in average crop.
Nitrate of Potash	I.	20.70	6.20	...26.71	0.221
	II.	19.22	7.30		
Sal-ammoniac	I.	15.10	4.93	...18.83	0.142
	II.	17.34	3.54		
Nitrate of ammonia	I.	12.29	3.72	...18.32	0.133
	II.	14.87	5.86		
Phosphate of ammonia	I.	12.96	3.77	...18.40	0.133
	II.	15.82	4.34		

It is seen that the ammonia-salts gave about one-fourth less crop than the nitrate of potash. The potash doubtless contributed somewhat to this difference.

The author began some experiments on this point in 1861, which turned out unsatisfactorily on account of the want of light in the apartment. In a number of these, buckwheat, sown in a weathered feldspathic sand, was manured with equal quantities of nitrogen, potash, lime, phosphoric acid, sulphuric acid, and chlorine, the nitrogen being presented in one instance in form of nitrate of potash, in the others as an ammonia-salt—sulphate, muriate, phosphate, or oxalate.

Although the plants failed to mature, from the cause above mentioned, the experiments plainly indicated the inferiority of ammonia as compared with nitric acid.

Explanations of this fact are not difficult to suggest. The most reasonable one is, perhaps, to be found in the circumstance that clayey matters (which existed in the soil under consideration) "fix" ammonia, i. e., convert it into a comparatively insoluble compound, so that the plant may not be able to appropriate it all.

On the other hand, Hellriegel (*Ann. d. Landw.*, VII., 53, u. VIII., 119) got a better yield of clover in artificial soil with sulphate of ammonia and phosphate of ammonia than with nitrate of ammonia or nitrate of soda, the quantity of nitrogen being in all cases the same.

As Sachs and Knop developed the method of Water-Culture, it was found by the latter that ammonia-salts did not effectively replace nitrates. The same conclusion was arrived at by Stohmann, in 1861 and 1863 (*Henneberg's Journ.*, 1862, 1, and 1864, 65), and by Rautenberg and Kühn, in 1863 (*Henneberg's Journ.*, 1864, 107), who experimented with sal-ammoniac, as well as by Birner and Lucanus, in 1864 (*Vs. St.*, VIII., 152), who employed sulphate and phosphate of ammonia.

The cause of failure lay doubtless in the fact, first noticed by Kühn, that so soon as ammonia was taken up by the plant, the acid with which it was combined, becoming free, acted as a poison.

In 1866, Hampe (*Vs. St.*, IX., 165), using phosphate of ammonia as the single source of nitrogen, and taking care to keep the solution but faintly acid, obtained a maize-plant which had a dry weight of 18 grams, including 36 perfect seeds; no nitrates were formed in the solution.

The same summer Kühn (*Vs. St.*, IX., 167) produced two small maize-plants, one with phosphate, the other with sulphate of ammonia as the source of nitrogen, but his experiments were interrupted by excessive heat in the glass-house.

In 1866, Beyer (*Vs. St.*, IX., 480) also made trials on the growth of the oat-plant in a solution containing bicarbonate of ammonia. The plants vegetated, though poorly, and several blossomed and even produced a few seeds. Quite at the close of the experiments the plants suddenly began to grow, with formation of new shoots. Examination of the liquid showed that the ammonia had been almost completely converted into nitric acid, and the increased growth was obviously connected with this nitrification.

In 1867, Hampe (*Vs. St.*, X., 176) made new experiments with ammonia-salts, and obtained one maize-plant $2^1/_2$ ft. high, bearing 40 handsome seeds, and weighing, dry, $25^1/_2$ grams. In these trials the seedlings, at the time of unfolding the sixth or seventh leaf, after consuming the nutriment of the seeds, manifested remarkable symptoms of disturbed nutrition, growth being suppressed, and the foliage becoming yellow. After a week or two the plants recovered their green color, began to grow again, and preserved a healthy appearance until mature. Experiment demonstrated that this diseased state was not affected by the concentration of the nourishing solution, by the amount of free acid or of iron present, nor by the illumination. Hampe observed that from these trials it *seemed that the plants, while young, were unable to assimilate ammonia or did so with difficulty, but acquired the power with a certain age.*

In 1868, Wagner (*Vs. St.*, XI., 288) obtained exactly the same results as Hampe. He found also that a maize-seedling, allowed to vegetate for two weeks in an artificial soil, and then placed in the nutritive solution, with phosphate of ammonia as a source of nitrogen, grew normally, without any symptoms of disease. Wagner obtained one plant weighing, dry, 26'|₂ grams, and carrying 48 ripe seeds. In experiments with carbonate of ammonia, Wagner obtained the same negative result as Beyer had experienced in 1866.

Beyer reports (*Vs. St.*, XI., 267) that his attempts to nourish the oat-plant in solutions containing ammonia-salts as the single source of nitrogen invariably failed, although repeated through three summers, and varied in several ways. Even with solutions identical to those in which maize grew successfully for Hampe, the oat seedlings refused to increase notably in weight, every precaution that could be thought of being taken to provide favorable conditions. It is not impossible that all these failures to supply plants with nitrogen by the use of ammonia-salts depend not upon the incapacity of vegetation to assimilate ammonia, but upon other conditions, unfavorable to growth, which are inseparable from the methods of experiment. A plant growing in a solution or in pure quartz sand is in abnormal circumstances, in so far that neither of these media can exert absorbent power sufficient to remove from solution and make innocuous any substance which may be set free by the selective agency of the plant.

Further investigations must be awaited before this point can be definitely settled. It is, however, a matter of little practical importance, since ammonia is so sparsely supplied by nature, and the ammonia of fertilizers is almost invariably subjected to the conditions of speedy nitrification.

CHAPTER VI.

THE SOIL AS A SOURCE OF FOOD TO CROPS.—INGREDIENTS WHOSE ELEMENTS ARE DERIVED FROM ROCKS.

§ 1.

GENERAL VIEW OF THE CONSTITUTION OF THE SOIL AS RELATED TO VEGETABLE NUTRITION.

Inert, Active, and Reserve Matters.—In all cases the soil consists in great part of matters that are of no direct or present use in feeding the plant. The chemical nature of this inert portion may vary greatly without correspondingly influencing the fertility of the soil. Sand, either quartzose, calcareous, micaceous, feldspathic, hornblendic, or augitic; clay in its many varieties; chalk, ocher (oxide of iron), humus; in short, any porous or granular material that is insoluble and little alterable by weather, may constitute the mass of the soil. The physical and mechanical characters of the soil are chiefly influenced by those ingredients which preponderate in quantity. Hence Ville has quite appropriately designated them the "mechanical agents of the soil." They affect fertility principally as they relate the plant to moisture and to temperature. They also have an influence on crops by gradually assuming more active forms, and yielding nourishment as the result of chemical changes. In general, it is probable that 99 per cent and more of the soil, exclusive of water, does not in the slightest degree contribute directly to the support of the present vegetation of our ordinary field products.

The hay crop is one that takes up and removes from the soil the largest quantity of mineral matters (ash-ingredients), but even a cutting of 2½ tons of hay carries off no more than 400 lbs. per acre. From the data given on page 158, we may assume the weight of the soil upon an acre, taken to the depth of one foot, to be 4,000,000 lbs. The ash-ingredients of a heavy hay crop amount therefore to but one ten-thousandth of the soil, admitting the crop to be fed exclusively by the 12 inches next the surface. Accordingly no less than 100 full crops of hay would require to be taken off to consume one per cent of the weight of the soil to this depth. We confine our calculation to the ash-ingredients because we have learned that the atmosphere furnishes the main supply of the food from which the combustible part of the crop is organized. Should we spread out over the surface of an acre of rock 4,000,000 lbs. of the purest quartz sand, and sow the usual amount of seed upon it, maintaining it in the proper state of moisture, etc., we could not produce a crop; we could not even recover the seed. Such a soil would be sterile in the most emphatic sense. But should we incorporate with such a soil a few thousand lbs. of the mineral ingredients of agricultural plants, together with some nitrates in the appropriate combinations and proportions, we should bestow fertility upon it by this addition and be able to realize a crop. Should we add to our acre of pure quartz the ashes of a hay crop, 400 lbs., and a proper quantity of nitrate of potash, we might also realize a good crop, could we but ensure contact of the roots of the plants with all the added matters. But in this case the soil would be fertile for one crop only, and after the removal of the latter it would be as sterile as before. We gather, then, that there are three items to be regarded in the simplest view of the chemical composition of the soil, viz., the *inert mechanical basis*, the *presently available nutritive ingredients*, and the *reserve matters* from which the available ingredients are supplied as needed.

In a previous chapter we have traced the formation of the soil from rocks by the conjoint agencies of mechanical and chemical disintegration. It is the perpetual operation of these agencies, especially those of the chemical kind, which serves to maintain fertility. The fragments of rock, and the insoluble matters generally that exist in the soil, are constantly suffering decomposition, whereby the elements that feed vegetation become available. What, therefore, we have designated as the inert basis of the soil, is inert for the moment only. From it, by perpetual change, is preparing the available food of crops. Various attempts have been made to distinguish in fact between these three classes or conditions of soil-ingredients; but the distinction is to us one of idea only. We cannot realize their separation, nor can we even define their peculiar conditions. We are ignorant in great degree of the power of the roots of plants to imbibe their food; we are equally ignorant of the mode in which the elements of the soil are associated and combined; we have, too, a very imperfect knowledge of the chemical transformations and decompositions that occur within it. We cannot, therefore, dissect the soil and decide what and how much is immediately available, and what is not. Furthermore, the soil is chemically so complex, and its relations to the plant are so complicated by physical and physiological conditions, that we may, perhaps, never arrive at a clear and unconfused idea of the mode by which it nourishes a crop. Nevertheless, what we have attained of knowledge and insight in this direction is full of value and encouragement.

Deportment of the Soil towards Solvents.—When we

put a soil in contact with water, certain matters are dissolved in this liquid. It has been thought that the substances taken up by water at any moment are those which at that time represent the available plant-food. This notion was based upon the supposition that the plant cannot feed itself at the roots save by matters in solution. Since Liebig has brought into prominence the doctrine that roots are able to attack and dissolve the insoluble ingredients of the soil, this idea is generally regarded as no longer tenable.

Again, it has been taught that the reserve plant-food of the soil is represented by the matters which acids (hydrochloric or nitric acid) are capable of bringing into solution. This is true in a certain rough sense only. The action of hydrochloric or nitric acid is indeed analogous to that of carbonic acid, which is the natural solvent; but between the two there are great differences, independent of those of degree.

Although we have no means of learning with positive accuracy what is the condition of the insoluble ingredients of the soil as to present or remote availability, the deportment of the soil towards water and acids is highly instructive, and by its study we make some approach to the solution of this question.

Standards of Solubility.—Before proceeding to details, some words upon the limits of solubility and upon what is meant by soluble in water or in acids will be appropriate. The terms soluble and insoluble are to a great degree relative as applied to the ingredients of the soil. When it is affirmed that salt is soluble in water, and that glass is insoluble in that liquid, the meaning of the statement is plain; it is simply that salt is *readily recognized* to be soluble and that glass is not ordinarily perceived to dissolve. The statement that glass is insoluble is, however, only true when the *ordinary standards of solubility* are referred to. The glass bottle which may contain water for years without perceptibly yielding aught of its mass to the liquid, does, nevertheless, slowly dissolve. We may make its solubility perceptible by a simple expedient. Pulverize the bottle to the finest dust, and thus extend the surface of glass many thousand or million times; weigh the glass-powder accurately, then agitate it for a few minutes with water, remove the liquid, dry and weigh the glass again. We shall thus find that the glass has lost several per cent of its original weight (Pelouze), and by evaporating the water, it will leave a solid residue equal in weight to the loss experienced by the glass.

§ 2.

AQUEOUS SOLUTION OF THE SOIL.

The soil and the rocks from which it is formed would commonly be spoken of as insoluble in water. They are, however, soluble to a slight extent, or rather, we should say, they contain soluble matters.

The quantity that water dissolves from a soil depends upon the amount of the liquid and the duration of its contact; it is therefore necessary, in order to estimate properly any statements respecting the solubility of the soil, to know the method and conditions of the experiment upon which such statements are based.

We subjoin the results of various investigations that exhibit the general nature and amount of matters soluble in water.

In 1852 Verdeil and Risler examined 10 soils from the grounds of the *Institut Agronomique*, at Versailles. In each case about 22 lbs. of the fine earth were mixed with pure lukewarm water to the consistence of a thin pap, and after standing several hours with frequent agitation the water was poured off; this process was repeated to the third time. The clear, faintly yellow solutions thus obtained were evaporated to dryness, and the residues were analyzed with results as follows, per cent·

Name of Field, etc.	Volatile & Organic Matter.	Ash.	Per cent of Ash.								
			Sulphate of Lime.	Carbonate of Lime.	Phosphate of Lime.	Oxide of Iron.	Alumina.	Chlorides of Sodium & Potas'm	Silica.	Potash & Soda.	Magnesia.
Mall...[Walk	43.00	57.00	48.92	25.60	4.27	1.55	0.62	7.63	5.49	3.77	—
Pheasant	70.50	29.90	31.49	35.29	2.16	0.47	trace	3.55	13.67	4.23	—
Turf	35.00	65.00	48.45	6.08	2.75	1.21	—	6.19	25.71	5.06	—
Queen's Ave..	44.00	56.00	43.75	6.08	6.32	2.00	trace	14.45	15.61	4.13	—
Kitchen Gard.	37.00	63.00	26.60	12.35	11.20	trace	trace	18.51	19.60	7.23	trace
Satory..[Galy	33.00	67.00	18.70	21.25	18.50	3.72	0.50	—	21.60	4.65	—
Clay soil of	48.00	52.00	18.75	45.61	3.83	0.95	1.55	9.14	5.00	7.60	7.60
Lime soil, do.	47.00	53.00	17.21	48.50	9.00	trace		6.21	5.50	—	8.32
Peat bog	46.00	54.00	24.43	30.61	0.92	5.15	trace	6.06	8.75	7.45	—
Sand pit	47.04	52.06	22.31	34.59	8.10	1.02	—	4.05	15.58	6.47	—

Here we notice that in almost every instance all the mineral ingredients of the plant were extracted from these soils by water. Only magnesia and chlorine are in any case missing. We are not informed, unfortunately, what amount of soluble matters was obtained in these experiments.

We next adduce a number of statements of the *proportion* of matters which water is capable of extracting from earth, statements derived from the analyses of soils of widely differing character and origin.

I. Very rich soil (excellent for clover) from St. Martin's, Upper Austria, treated with six times its quantity of cold water (Jarriges).

II. Excellent beet soil (but clover sick) from Schlanstaedt, Silesia, treated with 5 times its quantity of cold water (Jarriges).

III. Fair wheat soil, Seitendorf, Silesia, treated with 5 times its weight of cold water (Peters).

IV. Inferior wheat soil from Lampersdorf, Silesia—5-fold quantity of water (Peters).

V. Good wheat soil, Warwickshire, Scotland—10-fold quantity of hot water (Anderson).

VI. Garden soil, Cologne—3-fold amount of cold water (Grouven).

VII. Garden soil, Heidelberg—3-fold amount of cold water (Grouven).
VIII. Poor, sandy soil, Bickendorf—3-fold amount of cold water (Grouven).
IX. Clay soil, beet field, Liebesnitz, Bohemia, extracted with 9.6 times its weight of water (R. Hoffmann).
X. Peat, Meronitz, Bohemia, extracted with 16 times its weight of water (R. Hoffmann).
XI. Peaty soil of meadow, extracted with 8 times its weight of water (R. Hoffmann).
XII. Sandy soil, Moldau Valley, Bohemia, treated with twice its weight of water (R. Hoffmann).
XIII. Salt meadow, Stollhammer, Oldenburg (Harms).
XIV. Excellent beet soil, Magdeburg (Hellriegel).
XV. Poor beet soil, but good grain soil, Magdeburg (Hellriegel).
XVI. Experimental soil, Ida-Marienhütte, Silesia, treated with $2\frac{1}{2}$ times its weight of cold water (Küllenberg).
XVII. Soil from farm of Dr. Geo. B. Loring, Salem, Mass., treated with twice its weight of water (W. G. Mixter).

MATTERS DISSOLVED BY WATER FROM 100,000 PARTS OF VARIOUS SOILS.

	Lime.	Magnesia.	Potash.	Soda.	Phosphoric Acid.	Chlorine.	Sulphuric Acid.	Silica.	Oxide of Iron and Alumina.	Organic Matters.	Total.
I	18	2	13	8	2	1	5	11	5	53	134
II	5	2½	3	5½	trace	trace	trace	4½	6½	24	51
III	6	1	4	4	—	trace	1	2	2	23	43
IV	10	trace	1	2	—	trace	1	11	3	18	46
V	34	7	8	13	—	7	9	22	—	36	136
VI	17	3	9	7½	5	2½	6½	13½	1	22	87
VII	23	1½	7	4½	1½	1½	1	38	2	30	110
VIII	8	½	½	3½	trace	1	1½	20	—	10	45
IX	33½	3½	4½	9	5	3½	18	trace	—	70	147
X	164	11	47	12	trace	33	302	trace	77	449	1095
XI	92	44	21	24	trace	trace	11	1	2	230	425
XII	1	2½	2	1	trace	trace	trace	trace	—	33	39½
XIII	79	43	16	476	—	407	114	58	—	170	1393
XIV	19	3	3	5	1	4	4	20	3	88	150
XV	26	5	3	4	1	5	3	15	2	83	147
XVI	6½	2	1	3	½	5½	3½	12	7	12	53
XVII	8	2	6½	1	$1\frac{1}{10}$	7½		1½	17	12	55½

The foregoing analyses (all the author has access to that are sufficiently detailed for the purpose) indicate

1. That the quantity of soluble matters is greatest—400 to 1,400 in 100,000—in wet, peaty soils (X, XI, XIII), though their aqueous solutions are not rich in some of the most important kinds of plant-food, as, for example, phosphoric acid.

2. That poor, sandy soils (VIII, XII) yield to water the least amount of soluble matters,—40 to 45 in 100,000.

3. That very rich soils, and rich soils especially when recently and heavily manured as for the hop and beet crops (I, II, V, VI, VII, IX, XIV, XV, XVI), yield, in general, to water, a larger proportion of soluble matters than poor soils, the quantity ranging in the instances before us from 50 to 150 parts in 100,000.

4. It is seen that in most cases phosphoric acid is not present in the aqueous extract in quantity sufficient to be estimated; in some instances other substances, as magnesia, chlorine, and sulphuric acid, occur in traces only.

5. In a number of cases essential elements of plant-food, viz., phosphoric acid and sulphuric acid, are wanting, or their presence was overlooked by the analyst.

Composition of Drain-Water.—Before further discussion of the above data, additional evidence as to the kind and extent of aqueous action on the soil will be adduced. The water of rains, falling on the soil and slowly sinking through it, forms solutions on the grand scale, the study of which must be instructive. Such solutions are easily gathered in their full strength from the tiles of thoroughdrained fields, when, after a period of dry weather, a rainfall occurs, sufficient to saturate the ground.

Dr. E. Wolff, at Moeckern, Saxony, made two analyses of the water collected in the middle of May from newly laid tiles, when, after a period of no flow, the tiles had been running full for several hours in consequence of a heavy rain. The soil was of good quality. He found:

IN 100,000 PARTS OF DRAIN-WATER.

	Rye field.	Meadow.
Organic matters,	2.6	3.2
Carbonate of lime,	21.9	4.4
" " magnesia,	3.1	1.4
" " potash,	0.3	0.5
" " soda,	1.9	1.4
Chloride of sodium,	2.3	trace
Sulphate of potash,	1.2	trace
Alumina, } Oxide of iron, }	0.8	0.6
Silica,	0.7	0.4
Phosphoric acid,	trace	1.9
	34.8	13.8

Prof. Way has made a series of elaborate examinations on drain-waters furnished by Mr. Paine, of Farnham, Surrey. The waters were collected from the pipes (4–5 ft. deep) of thorough-drained fields in December, 1855, and in most cases were the *first flow* of the ditches after the autumn rains. The soils, with exception of 7 and 8, were but a few years before in an impoverished condition, but had been brought up to a high state of fertility by manuring and deep tillage. (*Jour. Roy. Ag. Soc.*, XVII, 133.)

IN 100,000 PARTS OF DRAIN-WATER.

	1 Wheat field.	2 Hop field.	3 Hop field.	4 Wheat field.	5 Wheat field.	6 Hop field.	7 Hop field.
Potash............	trace	trace	0.03	0.07	trace	0.31	trace
Soda.............	1.43	3.10	3.23	1.24	2.03	2.00	4.57
Lime.............	6.93	10.24	8.64	2.83	3.60	8.31	18.50
Magnesia.........	0.97	3.21	3.54	0.58	0.30	1.33	3.57
Oxide of iron and alumina.	0.59	0.07	0.14	none	1.85	0.50	0.71
Silica............	1.35	0.64	0.78	1.71	2.57	0.93	1.21
Chlorine..........	1.00	1.57	1.84	1.16	1.80	1.73	3.74
Sulphuric acid.....	2.35	7.85	6.23	2.44	1.84	4.45	13.58
Phosphoric acid....	trace	0.17	trace	trace	0.11	0.09	0.17
Nitric acid........	10.24	21.45	18.17	2.78	4.93	11.50	16.35
Ammonia.........	0.025	0.025	0.025	0.017	0.025	0.025	0.009
Soluble organic matter.	10.00	10.57	17.85	8.00	8.11	8.28	10.57
Total.............	34.885	58.095	60.525	21.227	27.195	39.455	72.979

Krocker has also published analyses of drain-waters collected in summer from poorer soils. He obtained

IN 100,000 PARTS:

	a	b	c	d	e	f
Organic matters,	2.5	2.4	1.6	0.6	6.3	5.6
Carbonate of lime,	8.4	8.4	12.7	7.9	7.1	8.4
Sulphate of lime,	20.8	21.0	11.4	1.7	7.7	7.2
Nitrate of lime,	0.2	0.2	0.1	0.2	0.2	0.2
Carbonate of magnesia,	7.0	6.9	4.7	2.7	2.7	1.6
Carbonate of iron,	0.4	0.4	0.4	0.2	0.2	0.1
Potash,	0.2	0.2	0.2	2.2	0.4	0.6
Soda,	1.1	1.5	1.3	1.0	0.5	0.4
Chloride of sodium,	0.8	0.8	0.7	0.3	0.1	0.1
Silica,	0.7	0.7	0.6	0.5	0.6	0.5
Total,	42.1	42.5	33.7	15.3	25.8	24.7

Krocker remarks (*Jour. für Prakt. Chem.*, 60–466) that phosphoric acid could be detected in all these waters, though its quantity was too small for estimation.

a and *b* are analyses of water from the same drains—*a* gathered April 1st, and *b* May 1st, 1853; *c* is from an adjoining field; *d*, from a field where the drains run constantly, where, accordingly, the drain-water is mixed with spring water; *e* and *f* are of water running from the surface of a field and gathered in the furrows.

Lysimeter-Water.—Entirely similar results were obtained by Zöller in the analysis of water which was collected in the Lysimeter of Fraas. The lysimeter* consists of a vessel with vertical sides and open above, the upper part of which contains a layer of soil (in these experiments 6 inches deep) supported by a perforated shelf, while below is a reservoir for the reception of water. The vessel is imbedded in the ground to within an inch of its upper edge, and is then filled from the diaphragm up with soil. In this condition it remains, the soil in it being exposed to the same influences as that of the field, while the water which percolates the soil gathers in the reservoir

---

* Measurer of solution.

below. Dr. Zöller analyzed the water that was thus collected from a number of soils at Munich, in the half year, April 7th to Oct. 7th, 1857. He found

IN 100,000 OF LYSIMETER-WATER:

Potash,	0.65	0.21	0.20	0.55	0.38
Soda,	0.71	0.56	0.74	2.37	0.00
Lime,	14.58	5.76	7.08	6.84	9.23
Magnesia,	2.05	0.89	0.13	0.29	0.51
Oxide of iron,	0.01	0.63	0.83	0.57	0.43
Chlorine,	5.75	0.95	2.08	3.91	3.53
Phosphoric acid,	0.22	—	—	—	—
Sulphuric acid,	1.75	2.71	2.78	2.93	3.35
Silica,	1.04	1.13	1.75	0.95	0.93
Organic matter, with some nitric and carbonic acids,	20.47	12.59	13.67	12.08	10.19
Total,	47.23	25.46	29.26	30.52	29.15

The foregoing analyses of drain and lysimeter-water exhibit a certain general agreement in their results. They agree, namely, in demonstrating the presence in the soil-water of all the mineral food of the plant, and while the figures for the total quantities of dissolved matters vary considerably, their average, 36½ parts to 100,000 of water, is probably about equally removed from the extremes met with on the one hand in the drainage from a very highly manured soil, and on the other hand in that where the soil-solution is diluted with rain or spring water.

It must not be forgotten that in the analyses of drainage water the figures refer to 100,000 parts of water; whereas in the analyses on p. 311, they refer to 100,000 parts of soil, and hence the two series of data cannot be directly compared and are not necessarily discrepant.

Is Soil-Water destitute of certain Nutritive Matters?
—We notice that in the natural solutions which flow off from the soil, phosphoric acid in nearly every case exists in quantity too minute for estimation; and when estimated, as has been done in a number of instances, its proportion does not reach 2 parts in 100,000. This fact, together with the non-appearance of the same substance and of oth-

er nutritive elements, viz., chlorine and sulphuric acid, in the Table, p.311, leads to the question, May not the aqueous solution of the soil be altogether lacking in some essential kinds of mineral plant-food in certain instances? May it not happen in case of a rather poor soil that it will support a moderate crop, and yet refuse to give up to water all the ingredients of that crop that are derived from the soil?

The weight of evidence supports the conclusion that water is capable of dissolving from the soil all the substances that it contains which serve as the food of plants. The absence of one or several substances in the analytical statement would seem to be no proof of their actual absence in the solution, but indicates simply that the substance was overlooked or was too small for estimation by the common methods of analysis in the quantity of solution which the experimenter had in hand. It would appear probable that by employing enough of the soil and enough water in extracting it, solutions would be easily obtained admitting of the detection and estimation of every ingredient. Knop, however, asserts (*Chem. Centralblatt*, 1864, 168) that he has repeatedly tested aqueous solutions of fruitful soils for phosphoric acid, employing the soils in quantities ranging from 2 to 22 lbs., and water in similar amounts, without in any case finding any traces of it. On the other hand Schulze mentions having invariably detected it in numerous trials; and Von Babo, in the examination of seven soils, found phosphoric acid in every instance but one, which, singularly enough, was that of a recently manured clay soil. In no case did he fail to detect lime, potash, soda, sulphuric acid, chlorine, and nitric acid; magnesia he did not look for. (*Hoffmann's Jahresbericht der Ag. Chem.*, I. 17.)

So Heiden, in answer to Knop's statement, found and estimated phosphoric acid in four instances in proportions ranging from 2 to 6 parts in 100,000 of soil. (*Jahresbericht der Ag. Chem.*, 1865, p. 34.)

It should be remarked that Knop's failure to find phosphoric acid may depend on the (uranium) method he employed, a method different from that commonly used.

Can the Soil-water supply Crops with Food? — Assuming, then, that all the soil-food for plants exists in solution in the water of the soil, the question arises, Does the water of the soil contain enough of these substances to nourish crops? In case of very fertile or highly manured fields, this question without doubt should be answered affirmatively. In respect of poor or ordinary soils, however, the answer has been for the most part of late years in the negative. While to decide such a question is, perhaps, impossible, a closer discussion of it may prove advantageous.

Russell (*Journal Highland and Ag. Soc.*, New Series, Vol. 8, p. 534) and Liebig (*Ann. d. Chem. u. Pharm.*, CV, 138) were the first to bring prominently forward the idea that crops are not fed simply from aqueous solutions. Dr. Anderson, of Glasgow, presents the argument as follows (his *Ag. Chemistry*, p. 113):

"In order to obtain an estimate of the quantity of the substances actually dissolved, we shall select the results obtained * by Way. The average rain-fall in Kent, where the waters he examined were obtained, is 25 inches. Now, it appears that about two-fifths of all the rain which falls escapes through the drains, and the rest is got rid of by evaporation.† An inch of rain falling on an English acre weighs rather more than a hundred tons; hence in the course of a year, there must pass off by the drains about 1,000 tons of drainage water, carrying with it, out of the reach of plants, such substances as it has dissolved, and

---

* On drain-waters, see p 313.

† From Parke's measurements, *Jour. Roy. Ag. Soc., Eng.*, Vol. XVII, p. 127.

1,500 tons must remain to give to the plant all that it holds in solution. These 1,500 tons of water must, if they have the same composition as that which escapes, contain only two and a half pounds of potash and less than a pound of ammonia. It may be alleged that the water which remains lying longer in contact with the soil may contain a larger quantity of matters in solution; but even admitting this to be the case, it cannot for a moment be supposed that they can ever amount to more than a very small fraction of what is required for a single crop."

The objection to this conclusion which Anderson alludes to above, but which he considers to be of little moment, is, perhaps, a serious one. The soil is saturated with water sufficiently to cause a flow from drains at a depth of 4 to 5 ft., for but a small part of the growing season. The Indian corn crop, for example, is planted in New England in the early part of June, and is harvested the first of October. During the four months of its growth, the average rain-fall is not enough to make a flow from drains for more, perhaps, than one day in seven. During six-sevenths of the time, then, there is a current of water ascending in the soil to supply the loss by evaporation at the surface. In this way the solution at the surface is concentrated by the carrying upward of dissolved matters. A heavy rain dilutes this solution, not having time to saturate itself before reaching the drains. Accordingly we find that the quantity of matters dissolved by water acting thoroughly on the surface soil is greater than that washed out by an equal amount of drain-water; at least such is the conclusion to be gathered from the experiments of Eichhorn and Wunder.

These chemists have examined the solution obtained by leaving soil in contact with *just sufficient water to saturate it* for a number of days or weeks. (*Vs. St.*, II, pp. 107–111.)

The soil examined by Eichhorn was from a garden near

Bonn, Prussia, not freshly manured, and was treated with about one-third its weight (36.5 per cent) of cold water for ten days.

Wunder employed soil from a field of the Experiment Station, Chemnitz, Saxony. This soil had not been recently manured, and was of rather inferior quality (yielded 15 bushels wheat per acre, English). It was also treated with about one-third its weight (34.5 per cent) of water for four weeks.

The solutions thus procured contained in 100,000 parts,

	Bonn.	Chemnitz.
Silica,	4.80	2.57
Sulphuric acid,	10.02	—
Phosphoric acid,	3.10	traces
Oxide of iron and alumina,	trace	1.17
Chloride of sodium,	5.86	4.76
Lime,	12.80	8.36
Magnesia,	3.84	3.74
Potash,	11.54	0.75
Soda,	1.10	3.04

If we assume with Anderson that 1,500 tons (= 3,360,000 lbs.) of water remain in these soils to feed a crop, and that this quantity makes solutions like those whose composition is given above, we have dissolved (in pounds per English acre) from the soil of

	Bonn.	Chemnitz.
Silica,	161	86
Sulphuric acid,	343	—
Phosphoric acid,	104	?
Oxide of iron and alumina,		39
Chloride of sodium,	197	160
Lime,	430	281
Magnesia,	129	126
Potash,	387	25
Soda,	37	102

These results differ widely from those based on the composition of drain-water. Eichhorn, by a similar calculation, was led to the conclusion that the soil he operated with was capable of nourishing the heaviest crops with its aqueous solution. Wunder, on the contrary, calculated that the Chemnitz soil yields insufficient matters for the ordinary amount of vegetation; and we see that as respects potash, the wants of grass and root crops could not be satisfied with the quantities in our computation, while sulphuric acid and phosphoric acid are nearly or entirely wanting. We do not, however, regard such calculations as decisive, either one way or the other. The quantity of water which may stand at the actual service of a crop is beyond our power to estimate with anything like certainty. Doubtless the amount assumed by Anderson is too large, and hence the calculations relative to the Bonn and Chemnitz soils *as above interpreted*, convey an exaggerated notion of the extent of solution.

Proper Concentration of Plant-Food. — Let us next inquire what *strength of solution* is necessary for the support of plants.

As has been shown by Nobbe (*Vs. St.*, VIII, p. 337), Birner & Lucanus (*Vs. St.*, VIII, p. 134), and Wolff (*Vs. St.*, VIII, p. 192), various agricultural plants flourish to extraordinary perfection when their roots are immersed in a solution containing about one part of ash-ingredients (together with nitrates) to 1,000 of water.

The solutions they employed contained the following substances in the proportions stated (approximately) below:

In 100,000 parts of Water.	Nobbe.	Birner & Lucanus.	Wolff.
Lime,	16	19	19
Magnesia,	3	6½	2½
Potash,	31	16	16
Phosphoric acid,	7	24	14
Chlorine,	21	none	2
Sulphuric acid,	6	13	4
Oxide of iron,	½	½	½
Nitric acid,	31¼	36	51
	116	115	109

Nobbe found further that the vigor of vegetation in his

solution was diminished either by reducing the proportion of solid matters below 0.5, or increasing it to 2 parts in 1,000 of water. The proper dilution of the food of plants for most vigorous growth and most perfect development is thus approximately indicated.

We notice, however, considerable latitude as regards the proportions of some of the most important ingredients which are usually present in least quantity in the aqueous solution of the soil. Thus, phosphoric acid in one case is thrice as abundant as in the other. We infer, therefore, that the minimum limit of the individual ingredients is not fixed by the above experiments, especially not for ordinary growth.

Birner and Lucanus communicate other results (*Vs. St.*, VIII., p. 154), which throw much light on the question under discussion. They compared the growth of the oat plant, when nourished respectively by a rich garden soil, by ordinary cultivated land, by a solution the composition of which is given above, and lastly by a natural aqueous solution of soil, viz., a *well-water*. Below is a statement of the weight in grams of an average plant, produced in these various media, as well as that of the grain yielded by it.

	Weight of average plant, dry.	Weight of dry Grain.	Dry crops compared with seed, the latter taken as unity.
Garden..........	5.27	1.23	193
Field............	1.75	0.63	64
Solution.........	3.75	1.53	137
Well-water.......	2.91	1.25	106

We gather from the above figures that well-water, in quantities of one quart for each plant, renewed weekly, gave a considerably heavier plant, straw, and grain, than a field under ordinary culture; the yield in *grain being double that of the latter*, and equal to that obtained in a rich garden soil.

The analysis of the well-water shows that the nutritive solution need not contain the food of plants in greater proportion than occurs in the aqueous extract of ordinary soils.

The well-water contained, in 100,000 parts,

Lime,	15.14
Magnesia,	1.53
Potash,	2.13
Phosphoric acid,	0.16
Sulphuric acid,	7.45
Nitric acid,	6.02

We thus have demonstration that a solution containing but one-and-a-half parts of phosphoric acid to ten million of water is competent, so far as this substance is concerned, to support a crop bearing twice as much grain as an ordinary soil could produce under the same circumstances of weather. Do we thus reach the limit of dilution? We cannot answer for agricultural plants, but in case of some other forms of vegetation, the reply is obvious and striking.

Various species of *Fucus*, *Laminaria*, and other marine plants, contain iodine in notable quantities. This element, so much used in photography and medicine, is made exclusively from the ashes of these sea-weeds, one establishment in Glasgow producing 35 tons of it annually. The iodine must be gathered from the water of the ocean in which these plants vegetate, and yet, although the starch-test is so delicate that one part of iodine can be detected when dissolved in 300,000 parts of water, it is not possible to recognize iodine in the " bitterns " which remain when sea-water is concentrated to the one-hundreth of its original bulk, so that its proportion must be less than one part in thirty millions of water! (*Otto's Lehrbuch der Chemie*, 4te, Aufl., pp. 743–4.)

Mode whereby dilute solutions may nourish Crops.— There are other considerations which may enable us to reconcile extreme dilution of the nutritive liquid of the soil, with the conveyance by it into the plant of the requisite quantity of its appropriate food. It is certain that the amount of matters found in solution at any given moment in the water of the soil by no means represents its power of supplying nourishment to vegetation.

If the water which has saturated itself with the soluble matters of the soil be deprived of a portion or all of these matters, as it might be by the absorptive action of the roots of a plant, the water would immediately act anew upon the soil, and in time would dissolve another similar quantity of the same substance or substances, and these being taken up by plants, it would again dissolve more, and so on as long and to such an extent as the soil itself would admit. In other words, the *same water* may act over and over again in the soil, to transfer from it to the crop the needful soluble matters. It has been shown that the substances dissolved in water may diffuse through animal and vegetable tissues independently of each other, and independently of the water itself. (H. C. G., p. 340.)

Deportment of the Soil to renewed portions of Water.—It remains to satisfy ourselves that the soil is capable of yielding soluble matters continuously to renewed portions of water. The only observations on this point that the writer is acquainted with are those made by Schulze and Ulbricht. Schulze experimented on a rich soil from Goldberg, in Mecklenburg (*Vs. St.*, VI., 411). This soil, in a quantity of 1,000 grams (= 2.2 lbs.) was slowly leached with pure water, so that one liter (= 1.056 quart) of liquid passed it in 24 hours. The extraction was continued during six successive days, and each portion was separately examined for total matters dissolved, and for phosphoric acid, which is, in general, the least soluble of the soil-ingredients.

The results were as follows, for 1,000 parts of extract,

Portion of aqueous extract.	Total matters dissolved.	Organic and volatile.	Inorganic.	Phosphoric acid.
1	0.535	0.340	0.195	0.0056
2	0.120	0.057	0.063	0.0082
3	0.261	0.101	0.160	0.0088
4	0.203	0.083	0.120	0.0075
5	0.260	0.082	0.178	0.0069
6	0.200	0.077	0.123	0.0044
	1.579	0.740	0.839	0.0414

We see that each successive extraction removed from the soil a scarcely diminished quantity of mineral matters, including phosphoric acid. In case of a poor soil, we should not expect results so striking, as regards quantity of dissolved matters, but doubtless they would be similar in kind.

This is shown by the investigations that follow.

Ulbricht gives (*Vs. St.*, V., 207) the results of the similar treatment of four soils. 1,000 grams of each were put in contact with four times as much pure water for three days, then two-thirds of the solution was poured off for analysis, and replaced by as much pure water; this was repeated ten times. Partial analyses were made of some of the extracts thus obtained; we subjoin the published results:

Dissolved by 40,000,000 parts of water from 1,000,000 parts of—

Loamy Sand from Heinsdorf.

	1st Extract.	2d Extract.	3d Extract.	4th Extract.	7th Extract.	10th Extract.
Potash............	30½	15	15	8		4
Soda..............	34	14	21	18		11
Lime..............	95	39	38	39		
Magnesia.........	30½	12	10	10		
Phosphoric acid...	trace.	1½	3	3		
Total.........	190	81½	87	78		

Loamy Sand from Wahlsdorf.

Potash..........	23	12	13	6		4
Soda............	26	16	20	16		6
Lime............	116	43	39	42	48	
Magnesia........	36½	15	14	12	14	
Phosphoric acid..	7	3	4	4		
Total.........	208½	89	90	80		

Loamy ferruginous Sand from Dahme, containing 4½ of humus.

Potash..........	7	6	7	7		3
Soda............	41	11	26	17		8
Lime............	96	70	55	48	62	
Magnesia........	14	10	9	7	8	
Phosphoric acid..	trace.	2	trace.	1		
Total.........	158	99	97	80		

Fine Sandy Loam from Falkenberg.

Potash..........	15	11	9	9
Soda............	47	12	12	8
Lime............	47	27	19	18
Magnesia........	17	8	5	6
Phosphoric acid..	3	2	trace.	trace.
Total.........	129	60	45	41

As Schulze remarks, it is practically impossible to exhaust a soil completely by water. This liquid will still dissolve something after the most prolonged or frequently renewed action, as not one of the components of the soil is possessed of absolute insolubility, although in a sterile soil the amount of matters taken up would presently become what the chemist terms "traces," or might be such at the outset.

The two analyses by Krocker, *a* and *b*, p. 314, made on water from the same drain, gathered at an interval of one month, further show that water, rapidly percolating the soil, continuously finds and takes up new portions of all its ingredients.

In addition to the simple solution of matters, the soil suffers constantly the chemical changes which have been already noticed, and are expressed by the term weathering. Matters insoluble in water to-day become soluble to-morrow, and substances that to-morrow resist the action of water are taken up the day after. In this way there is no limit to the solution of the soil, and we cannot therefore infer from what the soil yields to water at any given moment nor from what is taken out of it by any given amount of water, the real extent to which aqueous action operates, during the long period of vegetable growth, to present to the roots of a crop the indispensable ingredients of its food.

The discussion of the question as to the capacity of water to dissolve from the soil enough of the various ingredients to feed crops, while satisfactorily establishing this capacity in case of rich soils, and making evident that in poor soils most of the inorganic matters are presented to vegetation by water in sufficient quantity, does not entirely satisfy us in reference to some of the needful elements of the plant, especially phosphoric acid.

It is therefore appropriate, in this place, to pursue further inquiries into the mode by which vegetation acquires food from the soil, although to do so will somewhat interrupt the general plan of our chapter.

Direct action of Roots upon the Soil.—In noticing the means by which rocks are converted into soils, the action of the organic acids of the living plant has been mentioned. Since that chapter was written, further evidence has been obtained concerning the influence of the plant on the soil, which we now proceed to adduce.

Sachs (*Experimental Physiologie*, 189) gives an account of observations made by him on the action of roots on marble, dolomite (carbonate of lime and magnesia), magnesite (carbonate of magnesia), osteolite (phosphate of lime), gypsum, and glass. Polished plates of these substances were placed at the bottom of suitable vessels and covered several inches in depth with fine quartz sand. Seeds of various plants were planted in the sand and kept moist. The roots penetrated the sand and came in contact with the plates below, and branched horizontally on their surfaces. After several days or weeks the plates were removed and examined. The plants employed were the bean, maize, squash, and wheat. The carbonates of lime and magnesia and the phosphate of lime were plainly corroded where they had been in contact with the roots, so that the course of the latter could be traced without difficulty. Even the action of the root-hairs was manifest as a faint roughening of the surface of the stone either side of the path of the root. Gypsum and glass were not perceptibly acted on.

Dietrich has made a series of experiments (*Hoffmann's Jahresbericht*, VI, 3) on the amount of matters made soluble from basalt and sandstone, both coarsely powdered, and kept watered with equal quantities of distilled water, when supporting and when free from vegetation. The crushed rocks were employed in quantities of 9 and 11 lbs.; they were well washed before the trials with distilled water, and access of dust was prevented by a layer of cotton batting upon the surface. After removing the plants, at the termination of the experiments, each sample of rock-soil was washed with the same quantity of water, to which a hundredth of nitric acid had been added. It was found that the plants employed, especially lupins, peas, vetches, spurry, and buckwheat, assisted in the decomposition and solution of the basalt and sandstone. Not only did these plants take up mineral matters from the rock, but the latter contained besides, a larger amount of soluble matters than was found in the experiments where no plants were made to grow. The cereal grains had the same effect, but in less degree. In the subjoined table we give the total quantities of substances dissolved under the influence of the growing vegetation. These figures were obtained by adding to what was found in the washings of the rock-soils the ash of the crops, and subtracting from that sum the ash of the seeds, together with the matters made soluble in the same soils, which had sustained no plants, but which had been treated otherwise in a similar manner.

MATTERS DISSOLVED BY ACTION OF ROOTS.

	On 9 lbs. of sandstone.	On 11 lbs. of basalt.
Of 3 lupin plants	0.608 grams.	0.749 grams.
" 3 pea "	0.481 "	0.713 "
" 20 spurry "	0.268 "	0.365 "
" 10 buckwh't "	0.232 "	0.327 "
" 4 vetch "	0.221 "	0.251 "
" 8 wheat "	0.027 "	0.196 "
" 8 rye "	0.014 "	0.132 "

These trials appear to show conclusively that plants exert a decided effect on the soil. We are not informed, however, what particular substances are rendered soluble under this influence.

We conclude, then, that the direct action of the roots of a crop may in all cases contribute toward supplying it with food, and in many instances may be absolutely essential to its satisfactory growth.

Further Notice of Matters Soluble in Water.—The analyses we have quoted show that every chemical element of the soil may pass into aqueous solution. They also show that some substances are dissolved more easily and in greater quantity than others.

In general, *chlorine*, *nitric acid*, and *sulphuric acid*, are most readily and completely taken up by water, and, for the most part, in combination with *lime*, *soda*, and, *magnesia*. In some cases, sulphuric acid appears to exist in a difficultly soluble condition (*Van Bemmelen*, *Vs. St.*, VIII., 263).

Potash, *ammonia*, *oxide of iron*, *alumina*, *silica*, and *phosphoric acid*, are the substances which are usually soluble in but small proportion. These, together with lime, magnesia, and soda, it is difficult or impossible to wash out completely from a soil of good quality.

Very poor soils may be deficient in soluble forms of any or several of the above ingredients, and therefore readily admit of nearly complete extraction by a small amount of water.

Certain soils contain soluble salts of iron and alumina (sulphates and humates) in considerable quantity, and are for that reason unproductive. Such are many marsh lands, as well as upland soils containing bisulphide of iron (iron pyrites), of the kind that readily oxidizes to sulphate of protoxide of iron (copperas).

§ 3.

SOLUTION OF THE SOIL IN STRONG ACIDS.

The strong acids, hydrochloric (muriatic), nitric, and sulphuric, by virtue of their vigorous affinities, readily remove from the soil a considerable quantity of all its mineral ingredients. The quantity thus taken up is greatly more than can be dissolved in water, and is, in general, the greater, the more fertile the soil. Exceptions are soils consisting largely of carbonate of lime (chalk soils), or compounds of iron (ochreous soils). The different acids above named exercise very unlike solvent effects according to their concentration, the time of their action, the temperature at which they are applied, and the chemical nature and state of division of the soil.

The deportment of the minerals which chiefly constitute the soil towards these acids will enable us to understand their action upon the soil itself. Of these minerals quartz, feldspar, mica, hornblende, augite, talc, steatite, kaolinite, chrysolite, and chlorite, when not altered by weathering, nearly or altogether resist the action of even hot and moderately strong hydrochloric and nitric acids.

On the other hand, all carbonates, sulphates, and phosphates, are completely dissolved, while the zeolites and serpentine are decomposed, their alkalies, lime, etc., entering into solution, and the silica they contain separating, for the most part, as gelatinous hydrate.

According to the nature of the soil, and the concentration of the reagent, hydrochloric acid, the solvent usually employed, takes up from two to fifteen or more per cent.

Very dilute acids remove from the soil the bases, lime, magnesia, potash, and soda, in scarcely greater quantity than they are united with chlorine, and with sulphuric, phosphoric, carbonic, and nitric acids. Treatment with stronger acids takes up the bases above mentioned, particularly lime and magnesia, in greater proportion than the acids specified. We find that, by the stronger acids, *silica* is displaced from combination (and may be taken up by boiling the soil with solution of soda *after* treatment with the acid). It hence follows that silicates, such as are decomposable by acids, (zeolites) exist in the soil, although we cannot recognize them directly by inspection even with the help of the microscope. To this point we shall subsequently recur.

§ 4.

PORTION OF SOIL INSOLUBLE IN ACIDS.

When a soil has been boiled with concentrated hydrochloric acid for some time, or until this solvent exerts no further action, there may remain quartz, feldspar, mica, hornblende, augite, and kaolinite (clay), together with other similar silicates, which, in many cases, are ingredients of the soil. Treatment with concentrated sulphuric acid at very high temperatures (Mitscherlich), or syrupy phosphoric acid (A. Müller), decomposes all these minerals, quartz alone excepted. By making, therefore, in the first place, a mechanical analysis, as described on page 147, and subjecting the fine portion, which consists entirely or in great part of clay, to the action of these acids, the quantity of clay may be approximately estimated. Or, by melting the portion insoluble in acids with carbonate of soda, or acting upon it with hydrofluoric acid, the whole may be decomposed, and its elementary composition be ascertained by further analysis.

Notwithstanding an immense amount of labor has been expended in studying the composition of soils, and chiefly in ascertaining what and how much, acids dissolve from them, we have, unfortunately, very few results in the way of general principles that are of application, either to a scientific or a practical purpose. In a number of special cases, however, these investigations have proved exceedingly instructive and useful.

§ 5.

REACTIONS BY WHICH THE SOLUBILITY OF THE ELEMENTS OF THE SOIL IS ALTERED. SOLVENT EFFECT OF VARIOUS SUBSTANCES THAT ARE COMMONLY BROUGHT TO ACT UPON SOILS. THE ABSORPTIVE AND FIXING POWER OF SOILS.

Chemical Action in the Soil.—Chemistry has proved that the soil is by no means the inert thing it appears to be. It is not a passive jumble of rock-dust, out of which air and water extract the food of vegetation. It is not simply a stage on which the plant performs the drama of growth. It is, on the contrary, in itself, the theater of ceaseless activities; the seat of perpetual and complicated changes.

A large share of the rocks now accessible to our study at the earth's surface have once been soil, or in the condition of soil. Not only the immense masses of stratified limestones, sandstones, slates, and shales, that cover so large a part of the Middle States, but most of the rocks of New England have been soil, and have supported vegetable and animal life, as is proved by the fossil relics that have been disinterred from them.

We have explained the agencies, mechanical and chemical, by which our soils have been formed and are forming from the rocks. By a reverse metamorphosis, involving also the coöperation of mechanical and chemical and even of vital influences, the soils of earlier ages have been solidified and cemented to our rocks. Nor, indeed, is this process of rock-making brought to a conclusion. It is going on at the present day on a stupendous scale in various parts of the world, as the observations of geologists abundantly demonstrate. If we moisten sand with a solution of silicate of soda or silicate of potash, and then drench it with chloride of calcium, it shortly hardens to a rock-like mass, possessing enough firmness to answer many building purposes (Ransome's artificial stone). A mixture of lime, sand, and water, slowly acquires a similar hardness. Many clay-limestones yield, on calcination, a material (water-lime cement) which hardens speedily, even under water, and becomes, to all intents, a rock. Analogous changes proceed in the soil itself. Hard pan, which forms at the plow-sole in cultivated fields, and moor-bed pan, which makes a peat basin impervious to water in beds of sand and gravel, are of the same nature.

The bonds which hold together the elements of feldspar, of mica, of a zeolite, or of slate, may be indeed loosened and overcome by a superior force, but they are not destroyed, and reassert their power when the proper circumstances concur. The disintegration of rock into soil is, for the most part, a slow and unnoticed change. So, too, is the reversion of soil to rock, but it nevertheless goes on. The cultivable surface of the earth is, however, on the whole, far more favorable to disintegration than to petrifaction. Nevertheless, the chemical affinities and physical qualities that oppose disintegration are inherent in the soil, and constantly manifest themselves in the kind, if not in the degree, involved in the making of rocks. The fourteen elementary substances that exist in all soils are capable of forming and tend to form a multitude of combinations. In our enumeration of the minerals from which soils originate, we have instanced but a few, the more common of the many which may, in fact, contribute to its formation. The mineralogist counts by hundreds the natural compounds of these very elements, compounds which, from their capability of crystallization, occur in a visibly distinguishable shape. The chemist is able, by putting together these elements in different proportions, and under various circumstances, to identify a further number of their compounds, and both mineralogy and chemistry daily attest the discovery of new combinations of these same elements of the soil.

We cannot examine the soil directly for many of the substances which most certainly exist in it, on account of their being indistinguishable to the eye or other senses, even when assisted by the best instruments of vision. We have learned to infer their existence either from analogies with what is visibly revealed in other spheres of observation, or from the changes we are able to bring about and measure by the art of chemical analysis.

Absorptive Power of the Soil.—We have already drawn attention to the fact that various substances, when put in contact with the soil, in a state of solution in water, are withdrawn from the liquid and held by the soil. As has been mentioned on p. 175, the first appreciative record of this fact appears to have been published by Bronner, in 1836. In his work on Grape Culture occur the following passages: "Fill a bottle which has a small hole in its bottom with fine river sand or half-dry sifted garden earth. Pour gradually into the bottle thick and putrefied dung-liquor until its contents are saturated. The liquid that flows out at the lower opening appears almost odorless and colorless, and has entirely lost its original properties." After instancing the facts that wells situated near dung-pits are not spoiled by the latter, and that the foul water of the Seine at Paris becomes potable after filtering through porous sandstone, Bronner continues: "These examples sufficiently prove that the soil, even sand, possesses the property of attracting and fully absorbing the extractive matters *so that the water which subsequently passes is not able to remove them; even the soluble salts are absorbed, and are only washed out to a small extent by new quantities of water.*"

It was subsequently observed in the laboratory of Liebig, at Giessen, that water holding ammonia in solution, when poured upon clay, ran through deprived of this substance. Afterward, Messrs. Thompson and Huxtable, of England, repeated and extended the observations of Bronner, and in 1850, Professor Way, then chemist to the Roy. Ag. Soc. of Eng., published in the Journal of that Society, Vol. XI., pp. 313–379 an account of a most laborious and fruitful investigation of the subject. Since that time many chemists have studied the phenomena of absorption, and the results of these labors will be briefly stated in the paragraphs that follow.

There are two kinds of absorptive power exhibited by soils. One is purely physical, and is the consequence of adhesion or surface-attraction, exerted by the particles of certain ingredients of the soil. The other is a chemical action, and results from a play of affinities among certain of its components.

The physical absorptive power of various bodies, including the soil, has been already noticed in some detail (pp. 161–176). In experiments like those of Bronner, just alluded to, the absorption of the coloring and odorous ingredients of dung-liquor is doubtless a physical process. These substances are separated from solution by the soil just as a mass of clean wool separates indigo from the liquor of a dye-vat, or as bone-charcoal removes the brown color from syrup.

Chemical absorptions depend upon the formation of new compounds, and in many cases occasion chemical decompositions and displacements in such a manner that while one ingredient is absorbed, and becomes in a sense fixed, another is released from combination and becomes soluble. Brief notice has already been made of the chemical absorption of ammonia by the soil (p. 243). We shall now enter upon a fuller discussion of this and allied phenomena.

When solutions of the various soluble acids and bases existing in the soil, or of their salts, are put in contact with any ordinary earth for a short time, suitable examination proves that in most cases a chemical change takes place,—a reaction occurs between the soil and the substance.

If we provide a number of tall, narrow lamp-chimneys or similar tubes of glass, place on the flanged end of each a disk of cotton-batting, tying over it a piece of muslin, then support them vertically in clamps or by strings, and fill each of them compactly, two-thirds full of ordinary loamy soil, which should be free from lumps, we have an arrangement suitable for the study of the absorptive power in question.

Let now solutions, containing various soluble salts of the acids and bases existing in the soil, be prepared. These solutions should be quite dilute, but still admit of ready identification by their taste or by simple tests. We may employ, for example, any or all of the following compounds, viz., saltpeter, common salt, sulphate of magnesia, phosphate of soda, and silicate of soda.

If we pour solution of saltpeter on the soil, which should admit of its ready but not too rapid percolation, we shall find that the first portions of liquid which pass are no longer a solution of nitrate of potash, but one of nitrates of lime, magnesia, and soda. The potash has disappeared from solution* and become a constituent of the soil, while other bases, chiefly lime, have been displaced from the soil, and now exist in the solution with the nitric acid.

If we operate in a similar manner on a fresh tube of soil with solution of salt (chloride of sodium), we shall find by chemical examination that the soda of the salt is absorbed by the soil, while the chlorine passes through in combination with lime, magnesia, and potash. In case of sulphate of magnesia, magnesia is retained, and sulphates of lime, etc., pass through. With phosphates and silicates we find that not only the base, but also these acids are retained.

Law of Absorption and Displacement.—From a great number of experiments made by Way, Liebig, Brustlein, Henneberg and Stohmann, Rautenberg, Peters, Weinhold, Küllenberg, Heiden, Knop, and others, it is established as a general fact that all cultivable soils are able to decompose salts of the alkalies and alkali earths in a state of solution, in such a manner as to retain the base together with phosphoric and silicic acids, while chlorine, nitric acid, and sulphuric acid, remained dissolved, in union with some other base or bases besides the one with which they were originally combined. The absorptive power of the soil is, however, limited. After it has removed a certain quantity of potash, etc., from solution its action ceases, it has become saturated, and can take up no more. If, therefore, a large bulk of solution be filtered through a small volume of earth, the liquid, after a time, passes through unaltered.

---

* The absence of potash may be shown by aid of strong, cold solution of *tartaric acid*, which will precipitate bitartrate of potash (cream of tartar) from the original solution, if not too dilute, but not from that which has filtered through the soil. The presence of lime in the liquid that passes the soil may be shown by adding to it either carbonate or oxalate of ammonia.

Experiments to ascertain *how much* of a substance the soil is able to absorb are made by putting a known amount of the *dry* soil (e. g. 100 grms.) in a bottle with a given volume (e. g. 500 cubic cent.) of solution whose content of substance has been accurately determined. The solutions are most conveniently prepared so as to contain as many grms. of the salt to the liter of water as corresponds to the *atomic weight* or *equivalent* of the former, or one-half, one tenth, etc., of that amount. The soil and solution are kept in contact with occasional agitation for some hours or days, and then a measured portion of the liquid is filtered off and subjected to chemical analysis.

The absorptive power of the soil is exerted *unequally towards individual substances*. Thus, in Peters' experiments (*Vs. St.*, II., 140), the soil he operated with absorbed the bases in quantities diminishing in the following order:

Potash, Ammonia, Soda, Magnesia, Lime.

Another soil, experimented upon by Küllenberg (*Jahresbericht über Agricultur. Chemie*, 1865, p. 15), absorbed in a different order of quantity, as follows:

Ammonia, Potash, Magnesia, Lime, Soda.

As might be expected, *different soils exert absorptive power towards the same substance to an unequal extent*. Rautenberg (*Henneberg's Jour. für Landwirthschaft*, 1862, p. 62), operated with nine soils, 10,000 parts of which, under precisely similar circumstances, absorbed quantities of ammonia ranging from 7 to 25 parts.

The *time required for absorption* is usually short. Way found that in most cases the absorption of ammonia was complete in half an hour. Peters, however, observed that 48 hours were requisite for the saturation of the soil he employed with potash, and in the experiments of Henneberg and Stohmann (*Henneberg's Journal*, 1859, p. 35), phosphoric acid continued to be fixed after the expiration of 24 hours.

The *strength of the solution* influences the extent of absorption. The *stronger the solution*, the *more substance* is taken up from it by the soil. Thus, in Peters' experiments, 100 grms. of soil absorbed from 250 cubic centimeters of solutions of chloride of potassium of various degrees of concentration, as follows:

Strength of Solution.			Potash absorbed by		
Designation.	Quantity of potash in 250 c.c. of solution.		100 parts of soil.	By 10,000 parts in round numbers.	Proportion absorb'd
$1/80$ equiv.	=	0.1472 gram.	0.9888 gram.	10	$2/3$
$1/40$ "	=	0.2944 "	0.1381 "	14	$1/2$
$1/20$ "	=	0.5888 "	0.1990 "	20	$1/3$
$1/10$ "	=	1.1777 "	0.3124 "	31	$1/4$
$1/5$ "	=	2.3555 "	0.4503 "	45	$1/5$

A glance at the right-hand column shows that although *absolutely less* potash is absorbed from a weak solution than from a strong one, yet the weak solutions yield *relatively more* than those which are concentrated.

The quantity of base absorbed in a given time, also depends upon the *relative mass* of the solution and soil. In these experiments Peters treated a soil with various bulks of $1/20$ solution of chloride of potassium. The results are subjoined:—

From 250 c.c. of solution 10,000 parts of soil absorbed 20 parts.
"   500 "   "   "   "   "   "   "   25   "
"   1,000 "   "   "   "   "   "   "   29   "

The quantity of a substance absorbed by the soil depends somewhat on the *state of combination* it is in, i. e., on the substances with which it is associated. Peters found, for example, that 10,000 parts of soil absorbed from solutions of a number of potash-salts, each containing $1/20$ of an equivalent of that base expressed in grams, to the liter, the following quantities of potash:—

From phosphate,	49 parts.
" hydrate,	40 "
" carbonate,	32 "
" bicarbonate,	28 "
" nitrate,	25 "
" sulphate,	21 "
" chloride* and carbonate,	21 "
" chloride,	20 "

---

* Chloride of Potassium, KCl.

We observe that potash was absorbed in this case in largest proportion from the phosphate, and in least from the chloride. Henneberg and Stohmann, operating on a garden soil, observed a somewhat different deportment of it towards ammonia-salts. 10,000 parts of soil absorbed as follows:—

From	phosphate,	21	parts.
"	hydrate,	13	"
"	sulphate,	12	"
"	hydrate and chloride,*	$11^1|_2$	"
"	chloride,	11	"
"	nitrate,	11	"

Fixation neither complete nor permanent.—A point of the utmost importance is that none of the bases are ever *completely* absorbed even from the most dilute solutions. Liebig indeed, formerly believed that potash is entirely removed from its solutions. We find, in fact, that when a dilute solution of potash is slowly filtered through a large body of soil, the first portions contain so little of this substance as to give no indication to the usual tests. These portions are similar in composition to drain-waters, and like the latter they contain potash in very minute though appreciable quantity.

In accordance with the above fact, it is found that *water will dissolve and remove a portion* of the potash, etc., which a soil has absorbed.

Peters placed in 250 c.c. of a solution of chloride of potassium 100 grams of soil, which absorbed 0.2114 gram of potash. At the expiration of two days, one-half of the solution was removed, and its place was supplied with pure water. After two days more, one-half of the liquid was again removed, and an equal volume of water added;

---

* Chloride of Ammonium, NH_4Cl.

and this process was repeated ten times. The soil lost thus in the several washings as follows:

In 2d,	3d,	4th,	5th,	6th,	7th extract.
0.0075	0.0096	0.0082	0.0069	0.0075	0.0082 grams.
		In 8th,	9th,	10th extract.	
		0.0112	0.0201	0.083 grams.	

Removed in all, 0.0875 gram of potash.
Remained in soil, 0.1239 gram.

In these experiments one part of absorbed potash required 28,100 parts of water for solution.

Similar results were obtained by Henneberg and Stohmann with a soil which had absorbed ammonia; one part of this base required 10,000 parts of water for re-solution.

It has been already stated, *that the absorption of one base is accompanied by the liberation of a corresponding quantity of other bases, while the acid element, if it be sulphuric or nitric acid, or chlorine, is found in its original quantity in the solution.* As an illustration of this rule, the following data obtained by Weinhold in the treatment of a soil with sulphate of ammonia are adduced. The quantities are expressed in grams, except where otherwise stated.

Content of Solution

		before contact with the soil.		after contact with the soil.					
Liters of solution.	Amount of soil.	Sulphuric acid.	Ammonia.	Sulphuric acid.	Ammonia.	Potash.	Soda.	Lime.	Magnesia.
1	300	0.303	0.129	0.329	0.056	0.012	0.121	0.110	0.049
1½	200	0.455	0.193	0.488	0.120	0.011	0.034	0.105	0.030

We observe that the soil not only retained no sulphuric acid, but gave up a small quantity to the solution. Of the ammonia a little more than one-half in one case, and three-eighths in the other, was absorbed, and in the solution its place was supplied chiefly by lime, but to some extent also by potash, soda, and magnesia, which were dissolved from the soil. It is also to be noticed that in the two cases—unlike quantities of the same soil and solution having been employed—the bases were displaced in quantities that bear to each other no obvious relation.

Another fact which follows from the rule just illustrated, is the following: *Any base that has been absorbed by the soil, may be released from combination partly or entirely by any other.*

Peters subjected a soil which had been saturated with potash and subsequently washed copiously with water to the action of various solutions. The results, which exhibit the principle just stated, are subjoined. The soil was employed in portions of 100 grams, each of which contained 0.204 gram of absorbed potash. These were digested for three days with 250 c.c. of solutions (of nitrates) of the content below indicated.

For sake of comparison the amount of matters taken up by distilled water is added.

Content of solution.	Dissolved by the solution.					Absorbed by the soil.
	Lime.	Magnesia.	Potash.	Soda.	Ammonia.	
gram.						
0.2808 soda.	0.0671(?)	0.0006	0.0983	0.2197	——	0.0611 soda.
0.2165 ammonia.	0.0322	——	0.1455	0.0024	0.1596	0.0569 ammonia.
0.2996 lime.	0.2380	0.0020	0.1252	0.0252	——	0.0616 lime.
0.2317 magnesia.	0.0542	0.1726	0.1221	0.0245	——	0.0591 magnesia.
Dist. water.	trace	——	0.0434	0.0004	——	——

We notice that while distilled water dissolved about $^1/_4$ of the absorbed potash, the saline solutions took up two, three, or more times that quantity. We observe further that soda liberated lime and magnesia, ammonia liberated lime and soda, lime brought into solution magnesia and soda, and magnesia set free lime and soda from the soil itself.

Again, Way, Brustlein, and Peters, have shown in case of various soils they experimented with, that the *saturating of them with one base* (potash and lime were tried) *increases the absorbent power for other bases, and on the other hand, treatment with acids, which removes absorbed bases, diminishes their absorptive power.*

This fact is made evident by the following data furnished by Peters. The soils employed were

No. 1. Unaltered Soil.

No. 2. Soil heated with hydrochloric acid for some time, then thoroughly washed with water.

No. 3. No. 2, boiled with 10 grams of sulphate of lime and water, and washed.

No. 4. No. 2, boiled with solution of 10 grams of chloride of calcium, and well washed with water.

No. 5. No. 2, boiled with water and 10 grams of carbonate of lime.

No. 6. No. 2, boiled with solution of bicarbonate of lime, and washed.

Portions of 100 grams of each of the above were placed in contact with 250 c.c. of $\frac{1}{20}$ solution of chloride of potassium for three days. The results are subjoined:

Number of soil.	Dissolved by the solution.				Potash absorbed by the soil.
	Lime.	Magnesia.	Soda.	Chlorine.	
1......	0.0940	0.0084	0.0261	0.4482	0.1841
2......	0.0136		0.0004	0.4444	0.0227
3......	0.0784	0.0024	0.0019	0.4452	0.0882
4......	0.0560	0.0094	0.0024	0.4452	0.1243
5......	0.1175	0.0094	0.0019	0.4425	0.1378
6......	0.1456	0.0074		0.4404	0.2011

It is seen that the soil which had been washed with acid, absorbed but one-ninth as much as the unaltered earth. The treatment with the various lime-salts increased the absorbent power, in the order of the Table, until in the last instance it surpassed that of the original soil. Here, too, we observe that the absorption of potash accompanies and is made possible by the displacement of other bases, (in this case almost entirely lime, since the treatment with acid had nearly removed the others). We observe further that the quantity of chlorine remained the same throughout (within the limits of experimental error,) not being absorbed in any instance.

Way first showed that the absorptive power of the soil

Is diminished or even destroyed by *burning or calcination.* Peters, experimenting on this point, obtained the following results:

Potash absorbed from solution of chloride of potassium by

	unburned	burned
Vegetable mould,	0.2515	0.0202
Loam,	0.1841	0.1200

The Cause of the Absorptive Power of Soils for Bases when combined with chlorine, sulphuric, and nitric acids, has been the subject of several extensive investigations. Way, in his papers already referred to, was led to conclude that the quality in question belongs to some peculiar compound or compounds that are associated with the clayey or impalpable portion of the soil. That these bodies were compounds of the bases of the soil with silica, was a most probable and legitimate hypothesis, which he at once sought to test by experiment.

Various natural silicates, feldspars, and others, and some artificial preparations, were examined, but found to be destitute of action. Finally, a silicate of alumina and soda containing water was prepared, which possessed absorptive properties.

To produce this compound, pure alumina was dissolved in solution of caustic soda on the one hand, and pure silica in the same solution on the other. On mingling the two liquids, a white precipitate separated, which, when washed from soluble matters and dried at 212°, had the following composition [*]:

Silica,	.46.1
Alumina,	26.1
Soda,	15.8
Water,	12.0
	100.0

---

[*] Way gives the composition of the anhydrous salt, and says it contained, dried at 212°, *about* 12 *per cent of water.* In the above statement this water is included, since it is obviously an essential ingredient.

This compound is analogous in constitution to the zeolites, in so far as it is a highly basic silicate containing water, and is easy of decomposition. It is, in fact, decomposed by water alone, which removes from it silicate of soda, leaving insoluble silicate of alumina.

On digesting this soda-silicate of alumina with a solution of any salt of lime, Way found that it was decomposed, its soda was eliminated, and a *lime-silicate of alumina* was produced. In several instances he succeeded in replacing nearly all the soda by lime. *Potash-silicate of alumina* was procured by acting on either the soda or lime silicate with solution of a potash-salt; and, in a similar manner, *ammonia* and *magnesia-silicates* were generated. In case of the ammonia-compound, however, Way succeeded in replacing only about one-third of soda or other base by ammonia. All of these compounds, when acted upon by pure water, yielded small proportions of alkali to the latter, viz.:

The soda- silicate gave 3.36 parts of soda to 10,000 of water.
The potash- " " 2.27 " " potash " " " "
The ammonia- " " 1.06 " " ammonia " " " "

Way found furthermore that exposure to a strong heat destroyed the capacity of these substances to undergo the displacements we have mentioned.

From these facts Way, concluded that there exist in all cultivable soils, compounds similar to those he thus procured artificially, and that it is their presence which occasions the absorptions and displacements that have been noticed.

Way gives as characteristic of this class of double silicates, that there is a regular order in which the common bases replace each other. He arranges them in the following series:

Soda—Potash—Lime—Magnesia—Ammonia:

and according to him, potash can replace soda but not the other bases; while ammonia replaces them all: or each base replaces those ranged to its left in the above series, but none of those on its right. Way remarks, that "of course the reverse of this action cannot occur." Liebig (*Ann. der Chem. u. Pharm.*, xciv, 380) drew attention to the fact that Way himself in the preparation of the potash-alumina-silicate, demonstrated that there is no invariable order of decomposition. For, as he asserts, this compound may be obtained by digesting either the lime-alumina-silicate, or soda-alumina-silicate in nitrate or sulphate of potash, when the soda or *lime* is dissolved out and replaced by potash.

Way was doubtless led into the mistake of assuming a fixed order of replacements by considering these exchanges of bases as regulated after the ordinary manifestations of chemical affinity. His own experiments show that among these silicates there is not only no inflexible order of decomposition, but also no *complete* replacements.

The researches of Eichhorn, "Ueber die Einwirkung verdünnter Salzlösungen auf Ackererde," (*Landwirthschaftliches Centralblatt*, 1858, ii, 169, and *Pogg. Ann.*, No. 9, 1858), served to clear up the discrepancies of Way's investigation, and to confirm and explain his facts.

As Way's artificial silicates contained about 12 per cent of water, the happy thought occurred to Eichhorn to test the action of saline solutions on the hydrous silicates (zeolites) which occur in nature. He accordingly instituted some trials on chabazite, an abstract of which is here given.

On digesting finely pulverized chabazite (hydrous silicate of alumina and lime) with dilute solutions of chlorides of potassium, sodium, ammonium, lithium, barium, strontium, calcium, magnesium, and zinc, sulphate of magnesia, carbonates of soda and ammonia, and nitrate of cadmium, he found in every case that the basic element of these salts became a part of the silicate, while *lime* passed into the solution. The rapidity of the replacement varied exceedingly. The alkali-chlorides reacted evidently in two or three days. Chloride of barium and nitrate of cadmium were slower in their effect. Chlorides of zinc and strontium at first, appeared not to react; but after twelve days, lime was found in the solution. Chloride of magnesium was still tardier in replacing lime.

Four grams of powdered chabazite were digested with 4 grams of chloride of sodium and 400 cubic centimeters of water for 10 days. The composition of the original mineral (I,) and of the same after the action of chloride of sodium (II,) were as follows:

	I.	II.
Silica,	47.44	48.31
Alumina,	20.69	21.04
Lime,	10.37	6.65
Potash,	0.65	0.64
Soda,	0.42	5.40
Water,	20.18	18.33
Total,	99.75	100.37

Nearly one-half the lime of the original mineral was thus substituted by soda. A loss of water also occurred. The solution separated from the mineral, contained nothing but soda, lime, and chlorine, and the latter in precisely its original quantity.

By acting on chabazite with dilute chloride of ammonium (10 grams to 500 c.c. of water) for 10 days, the mineral was altered, and contained 3.33 per cent of ammonia. Digested 21 days, the mineral yielded 6.94 per cent of ammonia, and also lost water.

These ammonia-chabazites lost no ammonia at 212°, it escaped only when the heat was raised so high that water began to be expelled; treated with warm solution of potash it was immediately evolved. The ammonia-silicate was slightly soluble in water.

As in the instances above cited, there occurred but a partial displacement of lime. Eichhorn made corresponding trials with solutions of carbonates of soda and ammonia, in order to ascertain whether the formation of a soluble salt of the displaced base limited the reaction; but the results were substantially the same as before, as shown by analyzing the residue after removing carbonate of lime by digestion in dilute acetic acid.

Eichhorn found that the artificial soda-chabazite re-exchanged soda for lime when digested in a solution of chloride of calcium; in solution of chloride of potassium, both soda and lime were separated from it and replaced by potash. So, the ammonia-chabazite in solution of chloride of calcium, exchanged ammonia for lime, and in solutions of chlorides of potassium and sodium, both ammonia and lime passed into the liquid. The ammonia-chabazite in solution of sulphate of magnesia, lost ammonia but not lime, though doubtless the latter base would have been found in the liquid had the digestion been continued longer.

It thus appears that in the case of chabazite all the protoxide bases may mutually replace each other, time being the only element of difference in the reactions.

Similar observations were made with natrolite (hydrous silicate of alumina and soda,) as well as with chlorite and labradorite, although in case of the latter difficultly decomposable silicates, the action of saline solutions was very slow and incomplete.

Mulder has obtained similar displacements with the zeolitic minerals stilbite, thomsonite, and prehnite. (*Chemie der Ackerkrume*, I, 396). He has also artificially prepared hydrous silicates, having properties like those of Way, and has noticed that *sesquioxide of iron* readily participates in the displacements. Mulder also found that the gelatinous zeolitic precipitate obtained by dissolving hydraulic cement in hydrochloric acid, precipitating by ammonia and long washing with water, underwent the same substitutions when acted upon by saline solutions.

The precipitate he operated with, contained (water-free) in 100 parts:

```
Silica ............................................ 49.0
Alumina .......................................... 11.1
Oxide of iron .................................... 21.9
Lime .............................................. 6.9
Magnesia ......................................... 1.1
Insoluble matters with traces of alkalies, etc .. 10.0
                                                  ─────
                                                  100.0
```

On digesting portions of this substance with solutions of sulphates of soda, potash, magnesia, ammonia, for a single hour, all the lime was displaced and replaced by potash—two-thirds of it by soda and nearly four-fifths of it by magnesia and ammonia.

Further investigations by Rautenberg (*Henneberg's Jour. für Landwirthschaft*, 1862, pp. 405-454), and Knop (*Vs. St.*, VII, 57), which we have not space to recount fully, have demonstrated that of the bodies possible to exist in the soil, those in the following list do not possess the power of decomposing sulphates and nitrates of lime, potash, ammonia, etc., viz.:

Rautenberg.
- Quartz sand.
- Kaolinite (purified kaolin.)
- Carbonate of lime (chalk.)
- Humus (decayed wood.)
- Hydrated oxide of iron.
- Hydrated alumina.
- Humate of lime, magnesia, and alumina. } Knop.
- Phosphate of alumina.
- Gelatinous silica.
  "   " dried in the air.

These bodies have no absorptive effect, either separately or together.

These observers, together with Heiden (*Jahresbericht über Agriculturchemie*, 1864, p. 17), made experiments on soils to which hydrated silicates of alumina, and soda, or of lime, etc., were added, and found their absorptive power thereby increased.

Rautenberg and Heiden also found an obvious relation to subsist between the absorptive powers of a soil and certain of its ingredients. Rautenberg observed that the absorptive power of the nine soils he operated with was closely connected with the quantity of *alumina* and *ox-*

ide of iron which the soils yielded to hydrochloric acid. Heiden traced a similar relation between the silica set free by the action of acids on eleven soils and their absorptive power. Rautenberg and Heiden further confirmed what Way and Peters had previously shown, viz., that treatment of soil with acids diminished their absorbent power. These facts admit of interpretation as follows: Since neither silica, hydrated alumina, nor hydrated oxide of iron, *as such*, have any absorptive or decomposing power on sulphates, nitrates, etc., and since these bodies do not ordinarily exist as such to much extent in soils, therefore the connection found in twenty cases to subsist between their amount (soluble in acids) in the soil, and the absorptive power of the latter points to a compound of these (and other) substances (silicate of alumina, iron, lime, etc.), as the absorptive agent.

That the absorbing compound is not necessarily hydrated, is indicated by the fact that calcination, which must remove water, though it diminishes, does not always altogether destroy the absorptive quality of a soil. (See p. 343.) Eichhorn, as already stated, found that the anhydrous silicates, chlorite and labradorite, were acted upon by saline solutions, though but slowly.

Do Zeolitic Silicates, hydrated or otherwise, exist in the Soil?—When a soil which is free from carbonates and salts readily soluble in water, is treated with acetic, hydrochloric, or nitric acid, there is taken up a quantity (several per cent.) of matter which, while containing all the elements of the soil, consists chiefly of alumina and oxide of iron. Silica is not dissolved to much extent in the acid, but the soil which before treatment with acid contains but a minute amount of uncombined silica, afterwards yields to the proper solvent (hot solution of carbonate of soda) a considerable quantity. This is our best evidence of the presence in the soil of easily decomposable silicates. A number of analyses which illustrate these facts are subjoined:

		1. Sandy Loam. HEIDEN.	2.	3. White Clay.	4. Red Clay.	5. Porcelain Clay.	6. White Pottery Clay. WAY.
					RAUTENBERG.		
Water		1.613	1.347	6.15	6.39	10.36	6.18
Organic matter		2.387	2.003	none	none	none	none
Sand and insoluble silicates.		89.754	88.752	58.03	80.51	89.46	58.72
(Clay, kaolinite)		(10.344)	(5.762)				
Soluble in acids.	Silica	2.630*	4.199	18.73	6.80	0.04†	13.41
	Oxide of iron	1.872	1.630	2.11	0.90	0.14	5.38
	Alumina	1.152	1.288	12.15	4.35		13.90
	Lime	0.161	0.122	0.27	0.38	0.12	0.61
	Magnesia	0.201	0.210	0.29	0.17	0.08	0.43
	Potash	0.242	0.212	0.86			
	Soda	0.034	0.141	1.41			
	Phosphoric acid	0.083	0.034		0.50		1.37
	Sulphuric acid	0.007	0.021	none			
	Carbonic acid, chlorine, and loss	0.047	0.095	none			
		100.000	100.000	100.00	100.00	100.20	100.00

* This soil yielded to solution of carbonate of soda before treatment with acid, 0.340 % silica.
† The silica in this case is the small portion held in the acid solution.

The first three analyses especially, show that the soils to which they refer, contained a silicate or silicates in which iron, alumina, lime, magnesia and the alkalies existed as bases. How much of such silicates may occur in any given soil is impossible to decide in the present state of our knowledge. In the soil, free silica, is usually, if not always present, as may be shown by treatment with solution of carbonate of soda, but it appears difficult, if not impossible, to ascertain its quantity. Again, hydrated oxide of iron (according to A. Müller and Knop) and hydrated alumina* (Knop) may also exist, as can be made evident by digesting the soil in solution of tartrate of soda and potash (Müller, *Vs. St*, *IV, p*. 277), or in a mixture of tartrate and oxalate of ammonia (Knop, *Vs. St.* *VIII, p.* 41). Finally, organic acids occur to some extent in insoluble combinations with iron, alumina, lime,

* More probably, highly basic carbonates, or mixtures of hydrates and carbonates.

&c. This complexity of the soil effectually prevents an accurate analysis of its zeolitic silicates.

If further evidence of the existence of zeolitic compounds in the soil were needful, it is to be found in considering the analogy of the conditions which there obtain with those under which these compounds are positively known to be formed.

At Plombières, in France, the water of a hot spring (temperature, 140° F.) has flowed over and penetrated through a mass of concrete, composed of bricks and sandstone laid in lime, which was constructed centuries ago by the Romans. The water contains about nine ten-thousandths of solid matter in solution, a quantity so small as not to affect its taste perceptibly. As Daubrée has shown (*Ann. des Mines*, 5me., Série, T. XIII, p. 242), the cavities in the masonry frequently exhibit minute but well-defined crystals of various zeolitic minerals, viz.: chabasite, apophyllite, scolezite, harmotome, together with hydrated silicate of lime. These minerals have been produced by the action of the water upon the bricks and lime of the concrete, and while a high temperature prevails there, which probably has facilitated the crystallization of the minerals, as it certainly has done the chemical alteration of the bricks and sandstone, the conditions otherwise are just those of the soil.

In the soil, we should not expect to find zeolitic combinations crystallized or recognizable to the eye, because the small quantities of these substances that could be formed there must be distributed throughout twenty, fifty, or more times their weight of bulky matter, which would mechanically prevent their crystallization or segregation in any form, more especially as the access of water is very abundant; and the carbonic acid of the surface soil, which powerfully decomposes silicates, would operate antagonistically to their accumulation.

The water of the soil holds silica, lime, magnesia, alkalies, and oxide of iron, often alumina, in solution. Instances are numerous in which the evaporation of water containing dissolved salts has left a solid residue of silicates. Thus, Kersten has described (*Jour. für prakt. Chem.*, 22, 1) *a hydrous silicate of iron and manganese* that occurred as a hard incrustation upon the rock, in one of the Freiberg mines, and was deposited where the water leaked from the pumps. Kersten and Berzelius have noticed in the evaporation of mineral waters which contain carbonates of lime and magnesia, together with silica, that carbonates of these bases are first deposited, and finally silicates separate. (*Bischof's Chem. Geology, Car. Ed.*, Vol. 1, p. 5). Bischof (*loc. cit.*, p, 6) has found that silica, even in its most inactive form of quartz, slowly decomposes carbonate of soda and potash, forming silicate when boiled with their aqueous solutions. Undoubtedly, simple contact at ordinary temperature has the same effect, though more slowly and to a slight extent.

Such facts make evident that silica, lime, the alkalies, oxide of iron and alumina, when dissolved in water, if they do not already exist in combination in the water, easily combine when adverse affinities do not prevent, and may react upon the ingredients of the soil, or upon rock dust, with the formation of zeolites.

The "pan," which often forms an impervious stratum under peat bogs, though consisting largely of oxide of iron combined with organic acids, likewise contains considerable quantities of hydrated silicates, as shown by the analyses of Warnas and Michielsen (*Mulder's Chem. d. Ackerkrume,* Bd. 1, p. 566.)

Mulder found that when Portland cement (silicate of lime, alumina, iron, etc.) was treated with strong hydrochloric acid, whereby it was decomposed and in part dissolved, and then with ammonia, (which neutralized and removed the acid,) the gelatinous precipitate, consisting chiefly of free silica, free oxide of iron, free alumina, with smaller quantities of lime and magnesia, contained nevertheless a portion of silica, and of these bases in combination, because it exhibited absorbent power for bases, like Way's artificial silicates and like ordinary soil. Mere contact of soluble silica or silicates, with finely divided bases, *for a short time*, is thus proved to be sufficient for chemical union to take place between them.

Recently precipitated silicic acid being added to limewater, unites with and almost completely removes the lime from solution. The small portion of lime that remains in the liquid is combined with silica, the silicate not being entirely insoluble. (Gadolin, cited in *Storer's Dict. of Solubilities*, p. 551.)

The fact that *free bases*, as ammonia, potash and lime, are absorbed by and fixed in soils or clays that contain no organic acids, and to a degree different, usually greater than, when presented in combination, would indicate that they directly unite either with free silica or with simple silicates. The hydrated oxide of iron and alumina are indeed, under certain conditions, capable of retaining free alkalies, but only in minute quantities. (See p. 359.)

The fact that an admixture of carbonate of lime, or of other lime-salts with the soil, usually enhances its absorbent power, is not improbably due, as Rautenberg first suggested, to the formation of silicates.

A multitude of additional considerations from the history of silicates, especially from the chemistry of hydraulic cements and from geological metamorphism, might be adduced, were it needful to fortify our position.

Enough has been written, however, to make evident that *silica*, which is, so to speak, an accident in the plant, being unessential (we will not affirm useless) as one of its ingredients, is on account of its extraordinary capacity for chemical union with other bodies in a great variety of proportions, extremely important to the soil, and especially so when existing in combinations admitting of the remarkable changes which have come under our notice.

That we cannot decide as to the precise composition of the zeolitic compounds which may exist in the soil, is plain from what has been stated. We have the certainty of their analogy with the well-defined silicates of the mineralogist, which have been termed zeolites, an analogy of chemical composition and of chemical properties; we know further that they are likely to be numerous and to be in perpetual alteration, as they are subjected to the influence of one and another of the salts and substances that are brought into contact with them; but more than this, at present, we cannot be certain of.

Physical agencies in the phenomena of absorption.— While the absorption by the soil of potash or other base is accompanied by a chemical decomposition, which Way, Rautenberg, Heiden, and Knop's researches conclusively connect with certain hydrous silicates whose presence in the soil cannot be doubted, it has been the opinion of Liebig, Brustlein, Henneberg, Stohmann and Peters, that the real cause of the absorption is physical, and is due to simple surface attraction (adhesion) of the porous soil to the absorbed substance. Brustlein and Peters have shown that bone and wood-charcoal, washed with acids, absorb ammonia and potash from their salts to some extent, and after impregnation with carbonate of lime to as great an extent as ordinary soil. While the reasons already given appear to show satisfactorily that the absorbent power of the soil, *for bases in combination*, resides in the chemical action of zeolitic silicates, the facts just mentioned indicate that the physical properties of the soil may also exert an influence. Indeed, the fixation of *free bases* by the soil may be in all cases partially due to this cause, as Brustlein has made evident in case of ammonia (*Boussingault's Agronomie*, etc., T., II, p. 153).

Peters concludes the account of his valuable investigation with the following words: "*Absorption is caused by the surface attraction which the particles of earth exert. In the absorption of bases from salts, a chemical transposition with the ingredients of the soil is necessary, which is made possible through cöoperation of the surface attraction of the soil for the base.*" (Vs. St., II, p. 151.)

If we admit the soundness of this conclusion, we must also admit that in the soil the physical action is exerted *in sufficient intensity to decompose salts, by the hydrated silicates alone.* We must also allow that the displacements observed by Way and Eichhorn in silicates, are primarily due to mere physical action, though they have undeniably a chiefly chemical aspect.

That the phenomena are modified and limited in certain respects by physical conditions, is to be expected. The facts that the quantity of solution compared with the amount of soil, the strength of the solution, and up to a certain point the time of contact, influence the degree of absorption, point unmistakably to purely physical influences, analogous to those with whose action the chemist is familiar in his daily experience.

Absorption of Acids. — It has been mentioned already that phosphoric and silicic acids are absorbed by soils. Absorption of *phosphoric acid* has been invariably observed. In case of *silicic acid*, exceptions to the rule have been noticed. In very few instances has the absorption of sulphuric and nitric acids or chlorine, from their compounds, been remarked hitherto by those who have investigated the absorbent power of the soil. The nearly universal conclusion has been that these substances are not subject in any way, chemical or physical, to the attraction of the soil. Voelcker was the first to notice an absorption of *sulphuric acid* and *chlorine*. In his papers on "Farm Yard Manure," etc., (Jour. Roy. Ag. Soc., XVIII., p. 140,) and on the "Changes which Liquid Manure undergoes in contact with different Soils of Known Composition" (*idem* XX., 134–57), he found, in seven experiments, that dung liquor, after contact with various soils, lost or gained acid ingredients, as exhibited by the following figures, in grains per gallon : (loss is indicated by —, gain by +) :

	1	2	3	4	5	6	7	A.	B.
Chloride of Potassium	—8.81	+9.17	—2.71	+2.14	—2.71	+2.55	—1.10		
Chloride of Sodium	—3.95	—2.43	—7.01	—1.12	—1.10	—1.24	+3.66	—1.89	+19.05
Sulphuric Acid	+2.32	—4.21	—1.08	—1.21	—0.27	+1.21	+3.44	+2.26	—0.42
Silicic Acid	+1.63	+10.33	—1.61	+0.72	+2.76	—0.11	—0.07 undet.	—1.57	
Phosphoric Acid	—	—	—1.23	—3.09	—2.91	—3.38	—0.13	—8.76	—7.71

We notice that chlorine was perceptibly retained in three instances, while in the other four it was, on the whole, dissolved from the soil. Sulphuric acid was removed from the solution in four instances, and taken up by it in three others. In four cases silica was absorbed, and in three was dissolved. In his first paper, Professor Way recorded similar experiments, one with flax-steep liquor and a second with sewage. The results, as regards acid ingredients, are included in the above table, A and B, where we see that in one case a slight absorption of chlorine, and in the other of sulphuric acid, occurred. Way, however, regards these differences as due to the unavoidable errors of experiment, and it is certain that in Vœlcker's results similar allowance must be made. Nevertheless, these errors can hardly account for the large loss of chlorine observed in 1 and 3, or of sulphuric acid in 2.

Liebig found in his experiments "that a clay or lime-soil, poor in organic matter, withdrew from solution of silicate of potash, both silicic acid and potash, whereas one rich in humus extracted the potash, but left the silicic acid in solution." (Compare pp. 171–5.)

As regards *nitric acid*, Knop observed in a single instance that this body could not be wholly removed by water from a soil to which it had been added in known quantity. He regards it probable that it was actually retained rather than altered to ammonia or some other compound.

The fixation of acids in the soil is unquestionably, for the most part, a chemical process, and is due to the formation of comparatively insoluble compounds.

Hydrated oxide of iron and hydrated alumina are capable of forming highly insoluble compounds with all the mineral acids of the soil. The chemist has long been familiar with basic chlorides, nitrates, sulphates, silicates, phosphates and carbonates of these oxides. Whether such compounds can be actually produced in the soil is, however, to some extent, an open question, especially as regards chlorine, nitric and sulphuric acids. Their formation must also greatly depend upon what other substances are present. Thus, a soil rich in these hydrated oxides, and containing lime and the other bases in minuter quantity (except as firmly combined in form of silicates,) would not unlikely fix free nitric acid or free sulphuric acid as well as the chlorine of free hydrochloric acid. When the acids are presented in the form of salts, however, as is usually the case, the oxides in question have no power to displace them from these combinations. The acids, cannot, therefore, be converted into basic aluminous or iron salts unless they are first set free—unless the bases to which they were previously combined are first mastered by some separate agent. In the instance before referred to where nitric acid disappeared from a soil, Knop supposes that a basic nitrate of iron may have been formed, the soil employed being, in fact, highly ferruginous. The hydrated oxides of iron and alumina do, however, form insoluble compounds with *phosphoric acid*, and may even remove this acid from its soluble combinations with lime, as Thenard has shown, or even, perhaps, from its compounds with alkalies.

Phosphoric acid is fixed by the soil in various ways. When a phosphate of potash, for example, is put in contact with the soil, the base may be withdrawn by the absorbent silicate, and the acid may unite to lime or magnesia. The phosphates of lime and magnesia thus formed are, however, insoluble, and hence the acid as well as the base remains fixed. Again, if the alkali-phosphate be present in quantity so great that its base cannot all be taken up by the absorbent silicate, then the hydrated oxide of iron or alumina may react on the phosphate, chemically combining with the phosphoric acid, while the alkali gradually saturates itself with carbonic acid from the air. It is, however, more likely that organic salts of iron (crenates and apocrenates) transpose with the phosphate. So, too, carbonate of lime may decompose with phosphate of potash, producing carbonate of potash and phosphate of lime (J. Lawrence Smith). Vœlcker, in a number of experiments on the deportment of the soluble superphosphate of lime toward various soils, found that the absorption of phosphoric acid was more rapid and complete with soils containing much carbonate of lime than with clays or sands.

All observers agree that phosphoric acid is but slowly fixed by the soil. Vœlcker found the process was not completed in 26 days. Its absorption is, therefore, manifestly due to a different cause from that which completes the fixation of ammonia and potash in 48 hours.

As to *silicic acid*, it may also, as solid hydrate, unite slowly with the oxides of iron and with alumina (see Kersten's observations, p. 352). When occurring in solution, as silicate of an alkali, as happens in dung liquor, it would be fixed by contact with solid carbonate of lime, silicate of lime being formed (Fuchs, Kuhlmann), or by encountering an excess of solutions of any salt of lime, magnesia, iron or ammonia. In presence of free carbonic acid in excess, a carbonate of the alkali would be formed, and the silicic acid would be separated as such in a nearly insoluble form. Dung liquor, rich in carbonate of potash, on the other hand, would dissolve silica from the soil.

Sulphuric acid, existing in considerable quantities in dung liquor as a readily soluble salt of ammonia or potash, would be partially retained by a soil rich in carbonate of lime by conversion into sulphate of lime, which is comparatively insoluble.

Absorption of Bases, from their Hydrates, Carbonates and Silicates.—1. Incidentally it has been remarked that free bases, among which ammonia, potash, soda and lime are specially implied, may be retained by combining with undissolved *silica*. Potash, soda (and ammonia?) may at once form insoluble compounds if the silica be in large proportion; otherwise they may produce soluble silicates, which, however, in contact with lime, magnesia, alumina or iron salts, will yield insoluble combinations. As is well proved, gelatinous silica and lime at once form a nearly insoluble compound. It is probable that gelatinous silica may remove magnesia from solution of its bicarbonate, forming a nearly insoluble silicate of magnesia.

2. It has long been known that *hydrated oxide of iron* and *hydrated alumina* may unite with and retain free ammonia, potash, etc. Rautenberg experimented with both these substances as freshly prepared by artificial means, and found that, under similar conditions,

	10 grms. of hydrated oxide of iron.	10 grms. of hydrated alumina.
Absorbed of free ammonia	0.046 grm.	0.066 grm.
" " free potash	0.147 "	not det.

Long continued washing with water removes the alkali from these combinations. That oxide of iron and alumina commonly occur in the soil in quantity sufficient to have appreciable effect in absorbing free alkalies is extremely improbable.

Liebig has shown (*Ann. Ch. u. Ph.* 105, p. 122,) that hydrated alumina unites with silicate of potash with great avidity (an insoluble double silicate being formed just as in the experiments of Way, p. 343). According to Liebig, a quantity of hydrated alumina equivalent to 2.696 grms. of anhydrous alumina, absorbed from a liter of solution of silicate of potash containing 1.185 grm. of potash and 3.000 grm. of silica, fifteen per cent of the silicate. Doubtless hydrated oxide of iron would behave in a similar manner.

3. The *organic acids of humus* are usually the most effective agents in retaining the bases when the latter are in the free state, or exist as soluble carbonates or silicates. The properties of the humates have been detailed on page 230. It may be repeated here that they form with the alkalies* when the latter preponderate, soluble salts, but that these compounds unite readily to other earthy* and metallic* humates, forming insoluble compounds. Lime at once forms an insoluble humate, as do the metallic oxides. When, as naturally happens, the organic acids are in excess, their effect is in all cases to render the soluble free bases or their carbonates nearly insoluble.

In some cases, ammonia, potash and soda are absorbed more largely from their carbonates than from their hydrates. Thus, in some experiments made by the author, a sample of Peat from the New Haven Beaver Meadow was digested with diluted solution of ammonia for 48 hours, and then the excess of ammonia was distilled off at a boiling heat. The peat retained $0.95°|_0$ of this alkali. Another portion of the same peat was moistened with diluted solution of carbonate of ammonia and then dried at 212° until no ammoniacal smell was perceptible. This sample was found to have retained $1.30°|_0$ of ammonia. This difference was doubtless due to the fact that the

---

* In the customary language of Chemistry, potash, soda, and ammonia are alkalies or alkali-bases. Lime, magnesia, and alumina are earths or earthy bases, and oxide of iron and oxide of manganese are metallic bases.

peat contained *humate of lime*, which was not affected by the pure ammonia, but in contact with carbonate of ammonia yielded carbonate of lime and humate of ammonia. In these cases the ammonia was *in excess*, and the chemical changes were therefore, in some particulars, unlike those which occur when the humus preponderates.

Brustlein, Liebig and others have observed that soils rich in organic matter (forest mold, decayed wood,) have their absorptive power much enhanced by mixture with carbonate of lime.

Although Rautenberg has shown (*Henneberg's Journal* 186, p. 439,) that silicate of lime is probably formed when ordinary soils are mixed with carbonate of lime, it may easily happen, in the case of soils containing humus, that humate of lime is produced, which subsequently reacts upon the alkali-hydrates or salts with which absorption experiments are usually made.

§ 6.

REVIEW AND CONCLUSION.

The limits assigned to this work having been nearly reached, and the more important facts belonging to the present chapter brought under notice, with considerable fulness, it remains to sum up and also to adduce a few considerations which may appropriately close the volume. There are indeed a number of topics connected with the feeding of crops which have not been treated upon, such, especially as come up in agricultural practice; but these find their place most naturally and properly in a discussion of the improvement of the soil by tillage and fertilizers, to which it is proposed to devote a third volume.

What the Soil must contain.—In order to feed crops,

the soil must contain the ash-ingredients of plants, together with assimilable nitrogen-compounds in proper quantity and proportion. The composition of a very fertile soil is well exhibited by Baumhauer's analysis of an alluvial deposit from the waters of the Rhine, near the Zuider Zee, in Holland. This soil, which produces large crops, contained—

	Surface.	15 inches deep.	30 inches deep.
Insoluble silica, quartz,	57.646	51.706	55.372
Soluble silica,	2.340	2.496	2.286
Alumina,	1.830	2.900	2.888
Peroxide of iron,	9.039	10.305	11.864
Protoxide of iron,	0.350	0.563	0.200
Oxide of manganese,	0.288	0.354	0.281
Lime,	4.092	5.096	2.480
Magnesia,	0.130	0.140	0.128
Potash,	1.026	1.430	1.521
Soda,	1.972	2.069	1.937
Ammonia,*	0.060	0.078	0.075
Phosphoric acid,	0.466	0.324	0.478
Sulphuric acid,	0.896	1.104	0.576
Carbonic acid,	6.085	6.940	4.775
Chlorine,	1.240	1.302	1.418
Humic acid,	2.798	3.991	3.428
Crenic acid,	0.771	0.731	0.037
Apocrenic acid.	0.107	0.160	0.152
Other organic matters, and combined water (nitrates?),	8.324	7.700	9.348
Loss in analysis,	0.540	0.611	0.753
	100.000	100.000	100.000

A glance at the above analyses shows the unusual richness of this soil in all the elements of plant-food, with exception of nitrates, which were not separately determined. The alkalies, phosphoric acid, and sulphuric acid, were present in large proportion. The absolute quantities of the most important substances existing in an acre of this soil taken to the depth of one foot, and assuming this

---

* The figures are probably too high for ammonia, because, at the time the analyses were made, the methods of estimating this substance in the soil had not been studied sufficiently, and the ammonia obtained was doubtless derived in great part from the decomposition of humus under the action of an alkali.

quantity to weigh 3,500.000 lbs., (p. 158,) are as follows:

	lbs.
Soluble silica	81.900
Lime,	143.220
Potash,	35.910
Soda,	68,920
Ammonia,	2.100
Phosphoric acid	16.310
Sulphuric acid,	31.360
Nitric acid,	?

Quantity of Available Ash-ingredients necessary for a Maximum Crop.—We have already given some of the results of Hellriegel's experiments, made for the purpose of determining how much of the various elements of nutrition are required to produce a maximum yield of cereals (pp. 215 and 288). This experimenter found that 74 lbs. of nitrogen (in form of nitrates) to 1,000.000 of soil was sufficient to feed the heaviest growth of wheat. Of his experiments on the ash-ingredients of crops, only those relating to potash have been published. They are here reproduced.

EFFECTS OF VARIOUS PROPORTIONS OF AVAILABLE POTASH* IN THE SOIL ON THE BARLEY CROP.

Potash in 1,000.000 lbs. of soil.	Yield		
	of Straw and Chaff.	of Grain.	Total.
0	0.798		
6	3.809	2.993	6.802
12	5.740	4.695	10.435
24	6.859	7.851	14.710
47	8.195	9.578	17.773
71	9.327	10.097	19.424
94	8.693	9.083	17.776
141	8.764	8.529	17.293
282	8.916	8.962	17.878

It is seen that the greatest crop was obtained when 71 parts of potash were present in 1,000,000 lbs. of soil. A

---

* Other conditions were in all respects as nearly alike as possible.

larger quantity depressed the yield. It is probable that less than 71 lbs. would have produced an equal effect, since 47 lbs. gave so nearly the same result. The ash composition of barley, grain, and straw, in 100 parts, is as follows, according to Zoeller, (II. C. G., pp. 150 to 151):

	Grain.	Straw.
Potash,	18.5	12.0
Soda,	3.9	4.6
Magnesia,	7.0	3.0
Lime,	2.7	7.3
Oxide of iron,	0.7	1.9
Phosphoric acid,	32.4	6.0
Sulphuric acid,	2.8	2.8
Silica,	31.1	59.7
Chlorine,	1.1	2.6

The proportion of ash in the air-dry grain is $2\frac{1}{2}$ per cent, that in the straw is 5 per cent, (*Ann. Ch. u. Ph.* CXII, p. 40). Assuming the average barley crop to be 33 bushels of grain at 53 lbs. per bushel = 1,750 lbs., and one ton of straw,* we have in the barley crop of an acre the following quantities of ash-ingredients:

	Total ash-ingredients.	*Potash.*	*Soda.*	*Magnesia.*	*Lime.*	*Oxide of iron.*	*Phosphoric acid.*	*Sulphuric acid.*	*Chlorine.*
Barley Grain,	43.75	8.1	1.7	3.1	1.2	0.3	14.2	1.2	0.5
Straw,	100.00	12.0	4.6	3.0	7.3	1.9	6.0	2.8	2.6
Total,	143.75	20.1	6.3	6.1	8.5	2.2	20.4	4.0	3.1

In the account of Hellriegel's experiments, it is stated that the maximum barley crop in some other of his trials, corresponds to 8,160 lbs. of grain, or 154 bushels of 53 lbs. each per acre. This is more than $4\frac{1}{2}$ times the yield above assumed.

The above figures show that no essential ash-ingredient of the oat crop is present in larger quantity than potash. Phosphoric acid is quite the same in amount,

---

* These figures are employed by Anderson, and are based on Scotch statistics

while lime is but one-half as much, and the other acids and bases are still less abundant. It follows then that if 71 lbs. of available potash in 1,000.000 of soil are enough for a barley crop 4½ times greater than can ordinarily be produced under agricultural conditions, the same quantity of phosphoric acid, and less than half that amount of lime, etc., must be ample. Calculating on this basis, we give in the following statement the quantities required per acre, taken to the depth of one foot, to produce the maximum crop of Hellriegel (1), and the quantities needed for the average crop of 33 bushels (2). The amounts of nitrogen are those which Hellriegel found adequate to the wheat crop. See p. 289.

	1 lbs.	2 lbs.
Potash,	248	55
Soda,	78	17
Magnesia,	76	17
Lime,	105	23
Phosphoric acid,	250	55
Sulphuric acid,	49	11
Chlorine,	38	8
Nitrogen,	245	54

If now we divide the total quantities of potash, etc., found in an acre, or 3,500.000 lbs. of the soil analyzed by Baumhauer, by the number of pounds thus estimated to be necessarily present in order to produce a maximum or an average yield, we have the following quotients, which give the number of maximum barley crops and the number of average crops, for which the soil can furnish the respective materials.

The Zuider Zee soil contains enough

Lime	for 1364 maximum and 6138 average barley crops.
Potash	" 144 " " 648 " " "
Phosphoric acid	" 65 " " 292 " " "
Sulphuric	" " 64 " " 288 " " "
Nitrogen in ammonia	" 7 " " 31 " " "

We give next the composition of one of the excellent

wheat soils of Mid Lothian, analyzed by Dr. Anderson. The air-dry surface-soil contained in 100 parts:

Silica	71.552
Alumina	6.935
Peroxide of iron	5.173
Lime	1.229
Magnesia	1.082
Potash	0.354
Soda	0.433
Sulphuric acid	0.044
Phosphoric acid	0.430
Chlorine	traces
Organic matter	10.198
Water	2.684
	100.116

We observe that lime, potash, and sulphuric acid, are much less abundant than in the soil from the Zuider Zee. The quantity of phosphoric acid is about the same. The amount of sulphuric acid is but one-twentieth that in the Holland soil, and is accordingly enough for 15 good barley crops.

Lastly may be instanced the author's analysis of a soil from the Upper Palatinate, which was characterized by Dr. Sendtner, who collected it, as "the most sterile soil in Bavaria."

Water	0.535
Organic matter	1.850
Silica	0.016
Oxide of iron and alumina	1.640
Lime	0.096
Magnesia	trace
Carbonic acid	trace
Phosphoric acid	trace
Chlorine	trace
Alkalies	none
Quartz and insoluble silicates	95.863
	100.000

Here we note the absence in weighable quantity of magnesia and phosphoric acid, while potash could not even be detected by the tests employed. This soil was mostly naked and destitute of vegetation, and its composition shows the absence of any crop-producing power.

Relative Importance of the Ingredients of the Soil. —From the general point of view of vegetable nutrition, all those ingredients of the soil which act as food to the plant, are equally important as they are equally indispensable. Absence of any one of the substances which water-culture demonstrates must be presented to the roots of a plant so that it shall grow, is fatal to the productiveness of a soil.

Thus regarded, oxide of iron is as important as phosphoric acid, and chlorine (for the crops which require it) is no less valuable than potash. Practically, however, the relative importance of the nutritive elements is measured by their comparative abundance. Those which, like oxide of iron, are rarely deficient, are for that reason less prominent among the factors of a crop. If any single substance, be it phosphoric acid, or sulphuric acid, or potash, or magnesia, is lacking in a given soil at a certain time, that substance is then and for that soil the most important ingredient. From the point of view of natural abundance, we may safely state that, on the whole, available nitrogen and phosphoric acid are the most important ingredients of the soil, and potash, perhaps, takes the next rank. These are, most commonly, the substances whose absence or deficiency impairs fertility, and are those which, when added as fertilizers, produce the most frequent and remarkable increase of productiveness. In a multitude of special cases, however, sulphuric acid or lime, or magnesia, assumes the chief prominence, while in many instances it is scarcely possible to make out a greater crop-producing value for one of these substances over several others. Again, those ingredients of the soil which could be spared for all that they immediately contribute to the nourishment of crops, are often the chief factors of fertility on account of their indirect action, or because they supply some necessary physical conditions. Thus humus is not in any way essential to the growth of agricultural plants, for plants have been raised to full perfection without it; yet in the soil it has immense value practically, since among other reasons it stores and supplies water and assimilable nitrogen. Again, gravel may not be in any sense nutritious, yet because it acts as a reservoir of heat and promotes drainage it may be one of the most important components of a soil.

What the Soil must Supply.—It is not sufficient that the soil contain an adequate amount of the several ash-ingredients of the plant and of nitrogen, but it must be able to give these over to the plant in due quantity and proportion. The chemist could without difficulty compound an artificial soil that should include every element of plant-food in abundance, and yet be perfectly sterile. The potash of feldspar, the phosphoric acid of massive apatite, the nitrogen of peat, are nearly innutritious for crops on account of their immobility—because they are locked up in insoluble combinations.

Indications of Chemical Analysis.—The analyses by Baumhauer of soils from the Zuider Zee, p. 362, give in a single statement their ultimate composition. We are informed how much phosphoric acid, potash, magnesia, etc., exist in the soil, but get from the analysis no clue to the amount of any of these substances which is at the disposition of the present crop. Experience demonstrates the productiveness of the soil, and experience also shows that a soil of such composition is fertile; but the analysis does not necessarily give proof of the fact. A nearer approach to providing the data for estimating what a soil may supply to crops, is made by ascertaining what it will yield to acids.

Boussingault has analyzed in this manner a soil from Calvario, near Tacunga, in Equador, South America, which possesses extraordinary fertility.

He found its composition to be as follows:

Nitrogen in organic combination,	0.243
Nitric acid,	0.975
Ammonia,	0.010
Phosphoric acid, ⎫	0.400
Chlorine,	0.395
Sulphuric acid,	0.023
Carbonic acid, ⎬ Soluble in acids.	traces
Potash and Soda,	1.030
Lime,	1.256
Magnesia,	0.875
Sesquioxide of iron, ⎭	2.450
Sand, fragments of pumice, and clay insoluble in acids,	83.195
Moisture,	3.150
Organic matters (less nitrogen), undetermined substances, and loss,	5.938
	100.000

This analysis is much more complete in reference to nitrogen and its compounds, than those by Baumhauer already given (p. 362), and therefore has a peculiar value. As regards the other ingredients, we observe that phosphoric acid is present in about the same proportion; lime, alkalies, sulphuric acid, and chlorine, are less abundant, while magnesia is more abundant than in the soils from Zuider Zee.

The method of analysis is a guarantee that the one per cent of potash and soda does not exist in the insoluble form of feldspar. Boussingault found fragments of pumice by a microscopic examination. This rock is vesicular feldspar, or has at least a composition similar to feldspar, and the same insolubility in acids.

The inert nitrogen of the humus is discriminated from that which in the state of nitric acid is doubtless all assimilable, and that which, as ammonia, is probably so for the most part. The comparative solubility of the two per cent of lime and magnesia is also indicated by the analysis.

16*

Boussingault does not state the kind or concentration, or temperature of the acid employed to extract the soil for the above analysis. These are by no means points of indifference. Grouven (1ter & 3ter *Salzmünder Berichte*) has extracted the same earth with hydrochloric acid, concentrated and dilute, hot and cold, with greatly different results as was to be anticipated. In 1862, a sample from an experimental field at Salzmünde was treated, after being heated to redness, with boiling concentrated acid for 3 hours. In 1867 a sample was taken from a field 1,000 paces distant from the former, one portion of it was treated with boiling dilute acid (1 of concentrated acid to 20 of water) for 3 hours. Another portion was digested for three days with the same dilute acid, but without application of heat. In each case the same substances were extracted, but the quantities taken up were less, as the acid was weaker, or acted at a lower temperature. The following statement shows the composition of each extract, calculated on 100 parts of the soil.

EXTRACT OF SOIL OF SALZMÜNDE.

	Hot strong acid.	Hot dilute acid.	Cold dilute acid.
Potash,	.635	.116	.029
Soda,	.127	.067	.020
Lime,	1.677	1.046	1.098
Magnesia,	.687	.539	.237
Oxide of iron and alumina,	7.931	3.180	.650
Oxide of manganese,	.030	.086	.071
Sulphuric acid,	.059	.039	.020
Phosphoric acid,	.059	.091	.057
Silica,	1.785	.234	.175
Total,	12.990	5.398	2.357

The most interesting fact brought out by the above figures, is that strong and weak acids do not act on all the ingredients with the same relative power. Comparing the quantities found in the extract by cold dilute acid with those which the hot dilute acid took up, we find that the latter dissolved 5 times as much of oxide of iron and alumina, 4 times as much potash, 3 times as much soda, twice the amount of magnesia, sulphuric acid, and phosphoric acid, and the same quantity of lime. These facts show how very far chemical analysis in its present state is from being able to say definitely what any given soil can supply to crops, although we owe nearly all our precise knowledge of vegetable nutrition directly or indirectly to this art.

The solvent effect of water on the soil, and the direct action of roots, have been already discussed (pp. 309 to 328). It is unquestionably the fact that acids, like pure water in Ulbricht's experiments (p. 324), dissolve the more the longer they are in contact with a soil, and it is evident that the question: How much a particular soil is able to give to crops? is one for which we not only have no chemical answer at the present, but one that for many years, and, perhaps, always can be answered only by the method of experience—by appealing to the crop and not to the soil. Chemical analysis is competent to inform us very accurately as to the ultimate composition of the soil, but as regards its proximate composition or its chemical constitution, there remains a vast and difficult Unknown, which will yield only to very long and laborious investigation.

Maintenance of a Supply of Plant-food.—By the reciprocal action of the atmosphere and the soil, the latter keeps up its store of available nutritive matters. The difficultly soluble silicates slowly yield alkalies, lime, and magnesia, in soluble forms; the sulphides are converted into sulphates, and, generally, the minerals of the soil are disintegrated and fluxed under the influence of the oxygen, the water, the carbonic acid, and the nitric acid of the air, (pp. 122-135). Again, the atmospheric nitrogen is assimilated by the soil in the shape of ammonia, nitrates, and the amide-like matters of humus, (pp. 254-265).

The rate of disintegration as well as that of nitrification depends in part upon the chemical and physical characters of the soil, and partly upon temperature and meteorological conditions. In the tropics, both these processes go on more vigorously than in cold climates.

Every soil has a certain inherent capacity of production in general, which is chiefly governed by its power of supplying plant-food, and is designated its "natural strength." The rocky hill ranges of the Housatonic yield once in 30 years a crop of wood, the value of which, for a given locality and area, is nearly uniform from century to century. Under cultivation, the same uniformity of crop is seen when the conditions remain unchanged. Messrs. Lawes and Gilbert, in their valuable experiments, have obtained from "a soil of not more than average wheat-producing quality," without the application of any manure, 20 successive crops of wheat, the first of which was 15 bushels per acre, the last $17\frac{1}{2}$ bushels, and the average of all $16\frac{1}{4}$ bushels. (*Jour. Roy. Ag. Soc. of Eng.*, XXV, 490.) The same investigators also raised barley on the same field for 16 years, each year applying the same quantity and kinds of manure, and obtaining in the first 8 years (1852–59) an average of $44\frac{1}{5}$ bushels of grain and 28 cwt. of straw; for the second 8 years an average of $51\frac{3}{8}$ bushels of grain and 29 cwt. of straw; and for the 16 years an average of $48\frac{1}{5}$ bushels of grain and $28\frac{1}{2}$ cwt. of straw. (*Jour. of Bath and West of Eng. Ag. Soc.*, XVI, 244.)

The wheat experiments show the natural capacity of the Rothamstead soil for producing that cereal, and demonstrate that those matters which are annually removed by a crop of $16\frac{1}{4}$ bushels, are here restored to availability by weathering and nitrification. The crop is thus a measure of one or both of these processes.* It is probable

---

* In the experiments of Lawes and Gilbert it was found that phosphates, sulphates, and carbonates of lime, potash, magnesia, and soda, raised the produce of wheat but 2 to 3 bushels per acre above the yield of the unmanured soil, while sulphate and muriate of ammonia increased the crop 6 to 10 bushels. This result, obtained on three soils, viz., at Rothamstead in Herts, Holkham in Norfolk, and Rodmersham in Kent, the experiments extending over periods of 8, 3, and 4 years, respectively, shows that these soils were, for the wheat crop, relatively deficient in assimilable nitrogen. The crop on the unmanured soil was therefore a measure of nitrification rather than of mineral disintegration.

that this native power of producing wheat will last unimpaired for years, or, perhaps, centuries, provided the depth of the soil is sufficient. In time, however, the silicates and other compounds whose disintegration supplies alkalies, phosphates, etc., must become relatively less in quantity compared with the quite inert quartz and alumina-silicates which cannot in any way feed plants. Then the crop will fall off, and ultimately, if sufficient time be allowed, the soil will be reduced to sterility.

Other things being equal, this natural and durable productive power is of course greatest in those soils which contain and annually supply the largest proportions of plant-food from their entire mass, those which to the greatest extent originated from good soil-making materials.

Soils formed from nearly pure quartz, from mere chalk, or from serpentine (silicate of magnesia), are among those least capable of maintaining a supply of food to crops. These poor soils are often indeed fairly productive for a few years when first cleared from the forests or marshes; but this temporary fertility is due to a natural manuring, the accumulation of vegetable remains on the surface, which contains but enough nutriment for a few crops and wastes rapidly under tillage.

Exhaustion of the Soil in the language of Practice has a relative meaning, and signifies a reduction of producing power below the point of remuneration. A soil is said to be exhausted when the cost of cropping it is more than the crops are worth. In this sense the idea is very indefinite since a soil may refuse to grow one crop and yet may give good returns of another, and because a crop that remunerates in the vicinity of active demand for it, may be worthless at a little distance, on account of difficulties of transportation. The speedy and absolute exhaustion of a soil once fertile, that has been so much discussed by speculative writers, is found in their writings only, and does not exist in agriculture. A soil may be cropped below the point of remuneration, but the sterility thus induced is of a kind that easily yields to rest or other meliorating agencies, and is far from resembling in its permanence that which depends upon original poverty of constitution.

Significance of the Absorptive Quality.—Disintegration and nitrification would lead to a waste of the resources of fertility, were it not for the conserving effect of those physical absorptions and chemical combinations and replacements which have been described. The two least abundant ash-ingredients, viz., potash and phosphoric acid, if liberated by the weathering of the soil in the form of phosphate of potash, would suffer speedy removal did not the soil itself fix them both in combinations, which are at once so soluble that, while they best serve as plant-food, they cannot ordinarily accumulate in quantities destructive to vegetation, and so insoluble that the rain-fall cannot wash them off into the ocean.

The salts that are abundant in springs, rivers, and seas, are naturally enough those for which the soil has the least retention, viz., nitrates, carbonates, sulphates, and hydrochlorates of lime and soda.

The constituents of these salts are either required by vegetation in but small quantities as is the case with chlorine and soda, or they are generally speaking, abundant or abundantly formed in the soil, so that their removal does not immediately threaten the loss of productiveness. In fact, these more abundant matters aid in putting into circulation the scarcer and less soluble ingredients of crops, in accordance with the general law established by the researches of Way, Eichhorn, and others, to the effect that any base brought into the soil in form of a freely soluble salt, enters somewhat into nearly insoluble combination and liberates a corresponding quantity of other bases.

"The great beneficent law regulating these absorptions appears to admit of the following expression: *those bodies which are most rare and precious to the growing plant are by the soil converted into, and retained in, a condition not of absolute, but of relative insolubility, and are kept available to the plant by the continual circulation in the soil of the more abundant saline matters.*

"The soil (speaking in the widest sense) is then not only the ultimate exhaustless source of mineral (fixed) food, to vegetation, but it is the storehouse and conservatory of this food, protecting its own resources from waste and from too rapid use, and converting the highly soluble matters of animal exuviæ as well as of artificial refuse (manures) into permanent supplies."*

By absorption as well as by nitrification the soil acts therefore to prepare the food of the plant, and to present it in due kind and quantity.

---

* The author quotes here the concluding paragraphs of an article by him "On Some points of Agricultural Science," from the *American Journal of Science and Arts*, May, 1859. (p. 85), which have historic interest in being, so far as he is aware, the earliest, broad and accurate generalization on record, of the facts of soil-absorption.

NOTICE TO TEACHERS.

At the Author's request, Mr. Louis Stadtmuller, of New Haven, Conn., will undertake to furnish collections of the minerals and rocks which chiefly compose soils (see pp. 108–122), suitable for study and illustration, as also the apparatus and materials needful for the chemical experiments described in "How Crops Grow."

www.ingramcontent.com/pod-product-compliance
Lightning Source LLC
Chambersburg PA
CBHW020309240426
43673CB00039B/752